Musabekov Aydós

BIOLOGICAL STATISTICS

Musabekov Aydós

BIOLOGICAL STATISTICS

Training manual

ScienciaScripts

Imprint

Any brand names and product names mentioned in this book are subject to trademark, brand or patent protection and are trademarks or registered trademarks of their respective holders. The use of brand names, product names, common names, trade names, product descriptions etc. even without a particular marking in this work is in no way to be construed to mean that such names may be regarded as unrestricted in respect of trademark and brand protection legislation and could thus be used by anyone.

Cover image: www.ingimage.com

This book is a translation from the original published under ISBN 978-620-6-78087-8.

Publisher:
Sciencia Scripts
is a trademark of
Dodo Books Indian Ocean Ltd. and OmniScriptum S.R.L publishing group

120 High Road, East Finchley, London, N2 9ED, United Kingdom
Str. Armeneasca 28/1, office 1, Chisinau MD-2012, Republic of Moldova, Europe
Printed at: see last page
ISBN: 978-620-7-91272-8

Contents

The manual presents examples of solving typical problems faced by biologists and biotechnologists in the course of statistical processing of data obtained as a result of observations and experiments. The following methods are considered: data conformity to the law of normal distribution, sample size calculation, parametric and nonparametric analyses, multiple comparisons, factor analysis, correlation and regression analyses, etc. The algorithms of choice and step-by-step instructions for solving these problems using the STATISTICA software package are presented. The manual is intended for bachelors, masters and doctoral students of biological, biotechnological and agricultural areas of training.

Introduction

In recent years, various software tools for statistical analysis of data have become widespread. Despite this, the need to know at least the basics of mathematical statistics remains. A researcher should be able to choose appropriate statistical methods, know their capabilities and limitations, and interpret the results correctly and sensibly. Arbitrary application of even the most sophisticated methods of statistical analysis can lead to false conclusions.

In this regard, the aim of this manual is to develop an illustrative algorithm demonstrating the sequence of actions to be performed by a researcher when describing and analysing the results of a scientific study. In addition, some examples of various methods of statistical analysis implemented in the STATISTICA programme are considered.

This textbook summarises the whole range of mathematical statistical methods used in the collection and analysis of biological data. It provides examples of typical problems that biologists encounter in the course of statistical processing of data obtained from observations and experiments.

This manual can be used by bachelors, masters and doctoral students in the fields of training "Biology", "Biotechnology" and agricultural training, university teachers, all those interested in biological research and practical application of theoretical knowledge.

CHAPTER 1

Checking whether the analysed data conforms to the law of normality distributions.

The existing methods of statistical analysis can be divided into two large groups - parametric and nonparametric. An important condition determining the possibility of using a particular method of analysis is the subordination of the studied data to the law of normal (Gaussian) distribution, the graphical representation of which has the form of a characteristic bell-shaped curve (Fig. 2).

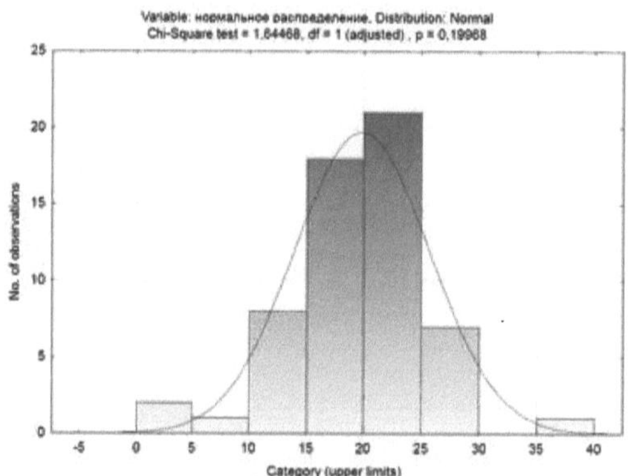

Figure 2. Example of "normal" (Gaussian) data distribution

If the data under study obey the law of normal distribution, parametric methods of analysis are used. Otherwise, non-parametric methods of statistical analysis are required. Application of parametric methods of analysis for data that do not obey the law of normal distribution of signs (distribution does not meet the criterion of "normality") leads to conclusions that do not correspond to reality.

It has been established that in the vast majority of cases (about 75%) the distribution of biological traits differs significantly from the "normal" distribution. In order to avoid the error mentioned above, the analysis of any biological data should start with checking the "normality" of their distribution.

Let us consider some approaches to assessing the "normality" of data distribution implemented in the STATISTICA programme.

Fig. 3 shows the results of counting the number of neuroglial cells in the substantia nigra of the rat brain. It is necessary to establish whether the distribution of these data obeys the law of normal distribution.

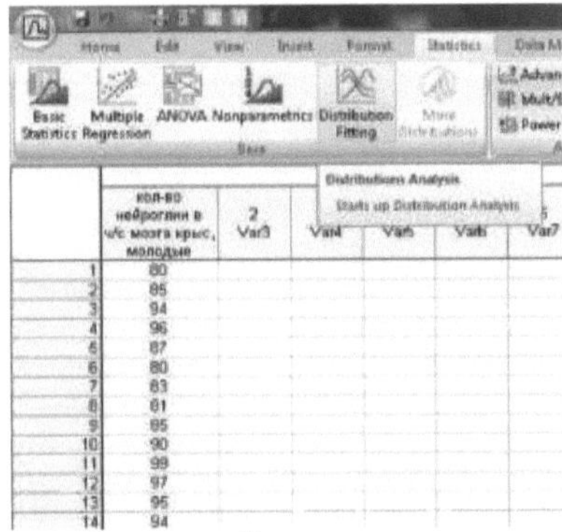

Figure 3. Data on the amount of rat neuroglia

An algorithm for selecting a method of statistical analysis.

The multitude of methods of mathematical statistics, complex description of procedures for their selection and implementation often confuse the researcher. However, a limited number of key factors are taken into account when selecting the necessary method of statistical analysis:

1. Type of data distribution. If the distribution of the data obtained in an experiment is considered as conforming to the law of normal distribution, parametric methods of analysis are used. For non-parametric methods of analysis, the type of data distribution does not matter.

2. The interconnectedness of the data under study. Interrelated (dependent) samples are those in which the characteristic under study is investigated on the same objects. If measurements of the studied attribute are carried out on different objects, the samples are considered as independent (non-interrelated). For mathematical processing of data in such problems, comparison methods for dependent or independent variables are used.

3. Quantitative characteristics of the data under study. If several factors influence the investigated characteristic, as well as when comparing several experimental groups, various types of multiple or variance analyses are used.

It is also important to be aware of the limitations of each type of statistical analysis. If the chosen method is not suitable for analysing the available data, some other method can always be found, perhaps by changing the type of presentation of the data itself.

Taking into account these factors, the algorithm for selecting a statistical analysis method can be presented in the form of the following scheme:

Checking experimental data for normality

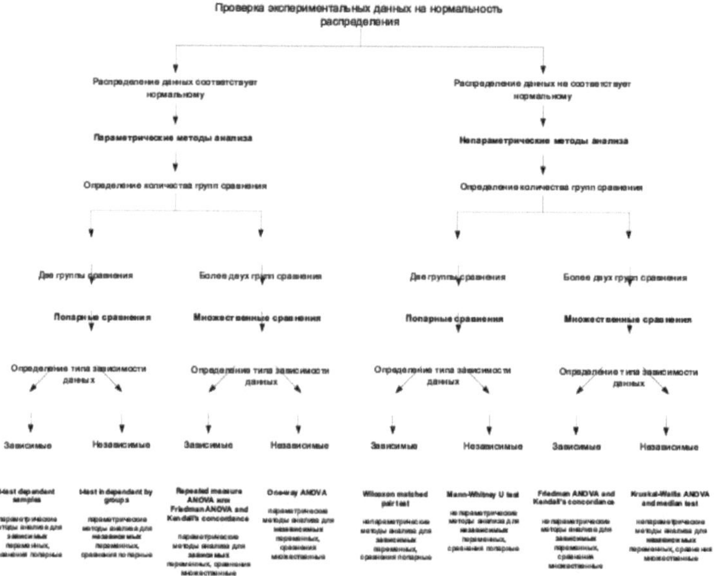

Fig. 1. Scheme (algorithm) of statistical analysis method selection for biological research

1. From the **Statistics** main menu section, launch a special module - **Distribution fitting**. This module allows you to check the data for compliance with a number of mathematical distributions (Fig. 3).
2. Since we need to check whether the data obey the law of normal distribution, select **Normal in** the list of **Continuous distributions** and click **OK** (Fig. 4).

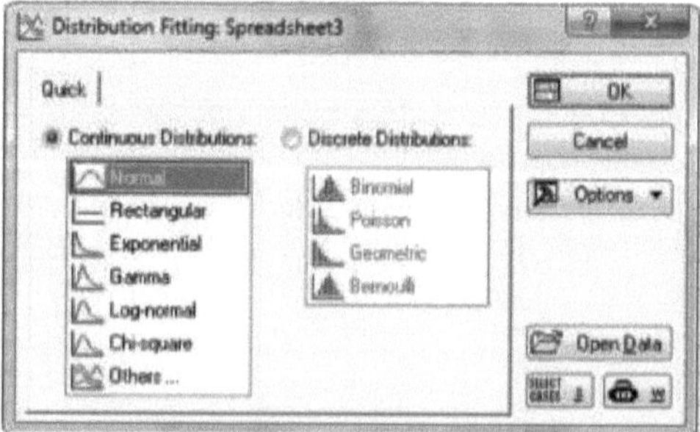

Fig. 4. Module dialogue window - Distribution fitting

3. In the next window, click on **the Variable** button, specify which variable we want to analyse. Then click the **Plot of observed and expected** distributions button (Figure 5).

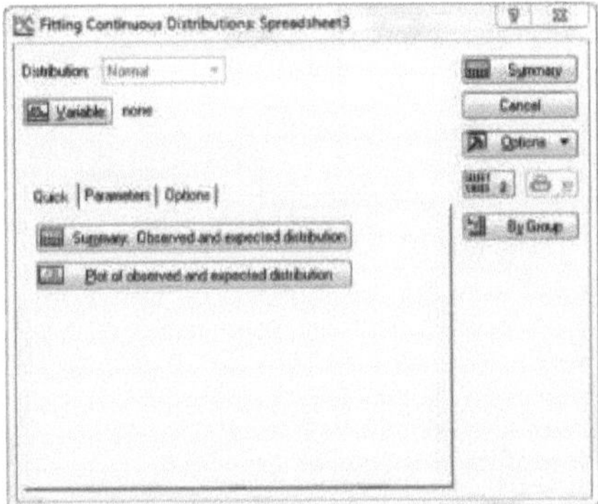

Figure 5. Dialogue window for selecting the variables to be analysed

The obtained histogram reflects the data distribution of the investigated parameter (Fig. 6).

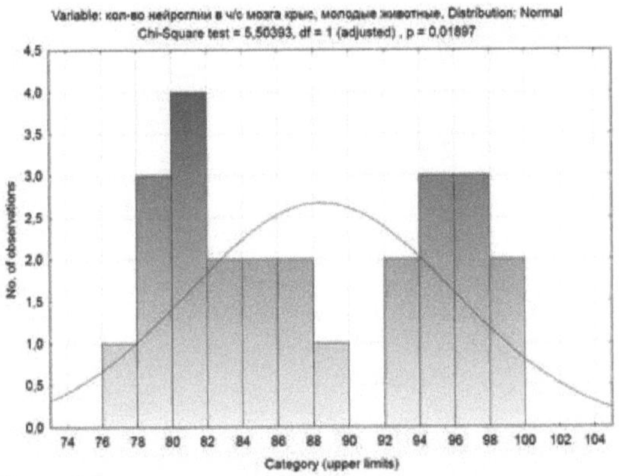

Figure 6. Result of analysing the distribution of the studied data

The obtained figure shows that the distribution of values of the studied parameter differs from "normal" (histogram bars do not form a bell-shaped curve). This conclusion is based on visual analysis, but it also has a more rigorous confirmation. The upper part of the histogram shows the results of the $\chi 2$ Chi-square test. This test tests the hypothesis that the observed distribution does not differ from the theoretically expected, "normal" distribution. If the probability of error in rejecting this hypothesis is much greater than 0.05 (p>0.05), then the hypothesis is true. In other words, the distribution of values constituting this sample is not statistically different from "normal". **In our case, the probability of error is less than 0.05 (p=0.01897), therefore, the distribution of values does not obey the "normal" law.**

However, it should be noted that the use of the chi-square test quite often leads to an erroneous conclusion about the "normality" of the distribution (the power of this test is relatively low). Therefore, it is better to use other tests, which can be <u>found in the **Basic** Statistics module </u>(Fig. 7).

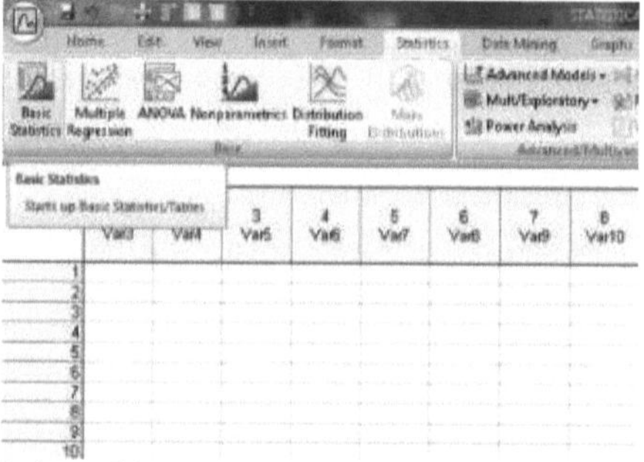

Fig. 7. The dialogue window of the Basic Statistics module

1. In the **Basic Statistics** section, select the **Descriptive Statistics** module (Figure 8).

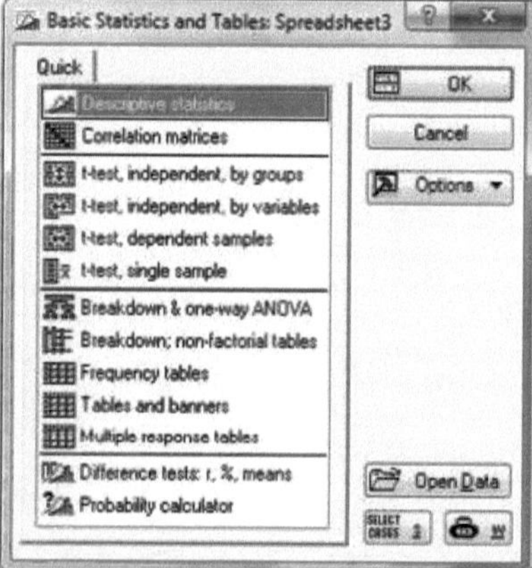

Fig. 8. The dialogue window of the Descriptive Statistics module

2. Open the **Normality** tab and select Kolmogorov-Smirnov **and Lilliefors test for normality** or **Shapiro-Wilk's W** test (Figure 9).

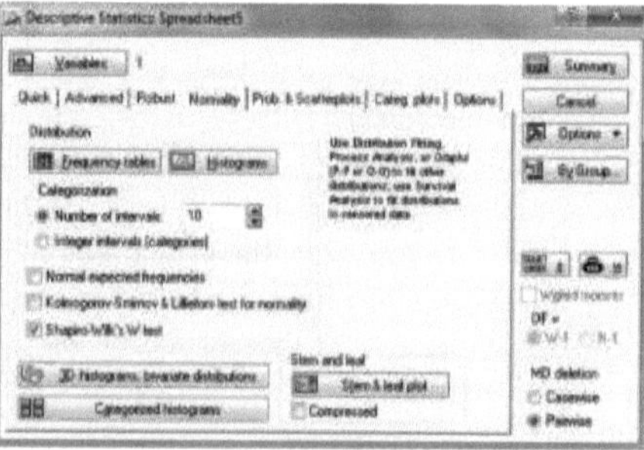

Fig. 9. The dialogue window of the Descriptive Statistics module

These tests also test the hypothesis that there is no difference between the observed distribution and the theoretically expected, "normal" distribution. The Shapiro-Wilk's W test has the greatest power, especially for small samples (n < 50). To select this test, you should tick the box next to its name.

3.   Next, click the **Variables button** and select the variable to be analysed. After clicking the **Histograms button**, the programme will create a histogram of the distribution of the trait values and the expected normal curve (Fig.10).

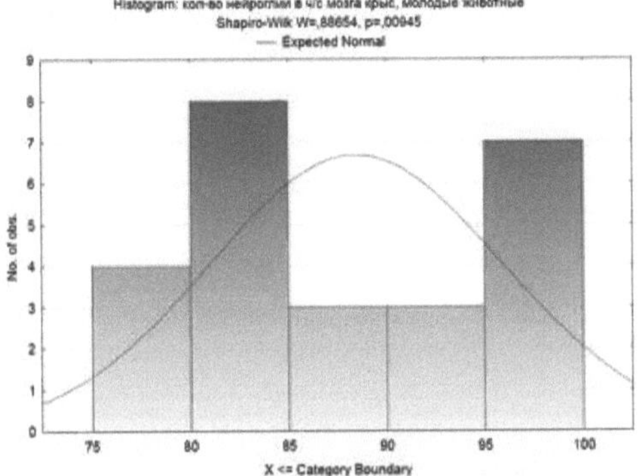

Figure 10. Result of analysing the distribution of the studied data

The results of the selected tests for "normality" are automatically placed in the header of this graph. In our example, the use of the Shapiro-Wilk test shows P = 0.00945, which confirms the earlier conclusion that the data do not obey the law of normal distribution.

You can also check the data for conformity to the law of normal distribution by using a normal probability plot. This graph shows the dependence of the actual frequencies of a trait value on the expected, "normal" frequencies. If there is no difference between the observed and expected distributions, the points on this graph will line up strictly along a straight line. Otherwise, they will form a figure different from a straight line. To draw a graph of this type, you need to:

1. From the main menu, select the **Basic Statistics** section and the **Descriptive Statistics** module (Figures 7 and 8).

2. In the dialogue box that appears, select the **Prob.** tab and click on the Normal probability plot button (Fig. 11). & Scatterplots tab and click on the **Normal probability plot** button (Fig. 11).

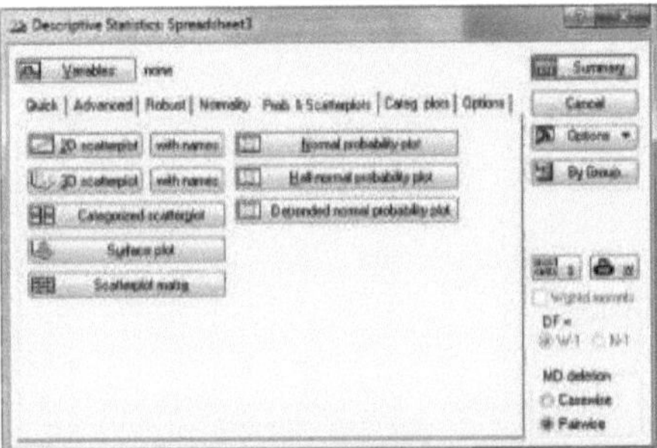

Fig. 11. The dialogue window of the Probability plots and scatter diagrams module

As a result, a graph will appear (Fig. 12), the points on which, in the case of a "normal" distribution of data, are densely lined up along the theoretically expected straight line. In our case, the points deviate significantly from the straight line, which once again confirms the assumption that the data do not conform to the law of normal distribution.

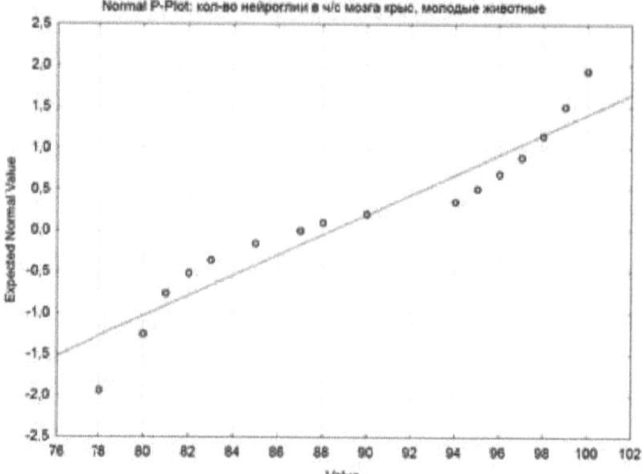

Fig. 12. Graph of normal probabilities

CHAPTER 2

A two-group comparison.

Comparison of two "independent" groups whose data distribution follows a "normal" distribution (T-test, independent, by groups).

In biological research, one of the most common tasks is to compare the arithmetic averages of two groups. An important characteristic of the groups being compared is their "dependence" or interrelationship.

Dependent samples contain data from studies of the same experimental group, but at different time periods. For example, "before" and "after" any impact: treatment (drug administration), education or training, surgery, etc. In this case, the measurement results obtained "before" and "after" the experimental impact will be interrelated with each other (interdependent). Note that the number of objects in these samples is always the same.

Independent samples are obtained when two different groups are studied. The results of a measurement in one sample do not affect the results obtained in the other sample. For example, 'experimental' and 'control' groups or comparisons between groups of men and women. It is acceptable to have different numbers of subjects.

A classic method for solving such a problem is Student's t-test, or simply "t-test". This test tests the hypothesis that the observed differences between the mean values of the compared samples are random and not caused by the factor under study (null hypothesis).

This test belongs to the group of parametric methods of analysis, its correct application requires fulfilment of three conditions:

1. The two samples should be independent;
2. Both samples must obey the law of normal distribution;
3. The two samples should be homogeneous (the spread of data within samples should not be too large).

The most important is the condition of compliance with the requirement to obey the law of normal distribution. **Failure to fulfil this requirement makes it impossible to apply this test.**

The algorithm for selecting this method of statistical analysis can be presented in the form of the following scheme (Fig. 13):

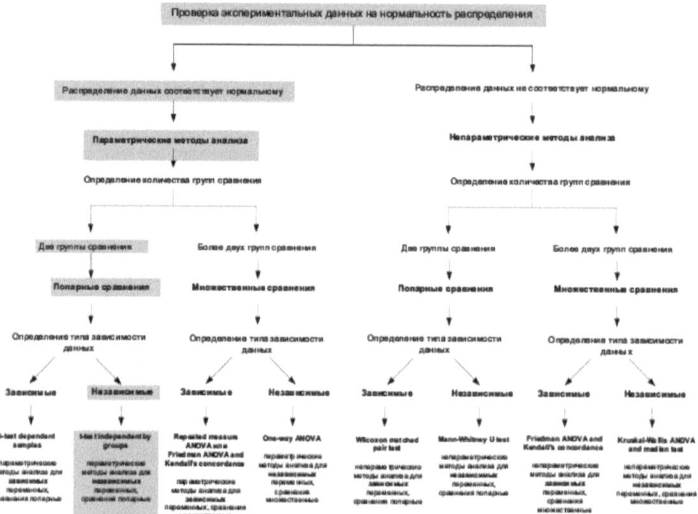

Fig. 13: Scheme for choosing the method of statistical analysis for comparing two independent groups, provided that the data conform to the law of normal distribution
Consider the application of the t-test using the following example.

It is known that aging processes are accompanied by the death of nerve cells in various parts of the brain. It is assumed that a possible cause of neuronal death is the increase of inflammatory processes in the nervous tissue. To assess the intensity of inflammation, we counted the number of glial cells (they are markers of inflammation) in animals (rats) of different age groups. Statistical analysis should answer the question whether the average number of these cells and, consequently, the intensity of inflammatory processes differ in animals of different ages.

Fig. 14 shows the data on the number of glial cells in two experimental groups: group 1 - young animals; group 2 old animals. **Note the design of the data (Fig. 14)** - the table has 2 variables. The first variable - **grouping variable** (Grouping variables) contains codes indicating that the data belong to a particular group. The second variable - Dependent variables - contains the data itself. The data themselves (in our case, the number of cells) are arranged vertically (one column below the other). The data are arranged in the same way in all cases when comparing independent data when comparing two independent groups

1 Grouping variable	2 Кол-во клеток глии
1	23
1	49
1	42
1	47
1	46
2	87
2	109
2	41
2	79
2	106
2	126
2	85
2	77

Figure 14: Example of data layout when comparing two independent groups

In the simplest version, the data for each group ("young" and "old") can simply be entered in separate columns, but when comparing independent groups, the first version of their design is preferable.

Assume that the data in both samples are "normally distributed" and the variances differ insignificantly. In order to perform the t-test it is necessary to:

1. Start the corresponding module from the menu: **Statistics / Basic statistics / t-test, independent, by groups** (Fig. 15).

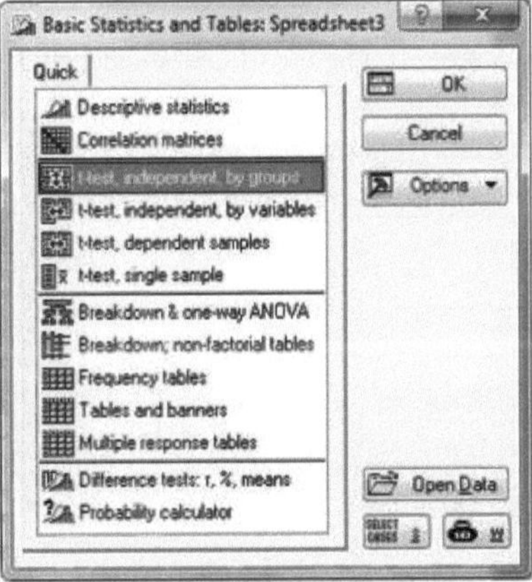

Fig. 15. Dialogue window of the Basic Statistics module

2. In the window that opens, specify the group codes in the corresponding boxes (**Code for Group 1 and 2**) or click the **Variables** button and specify the variables you want to use

In the right window it is necessary to select the group variable, in the left window - the dependent variable (Fig. 16).

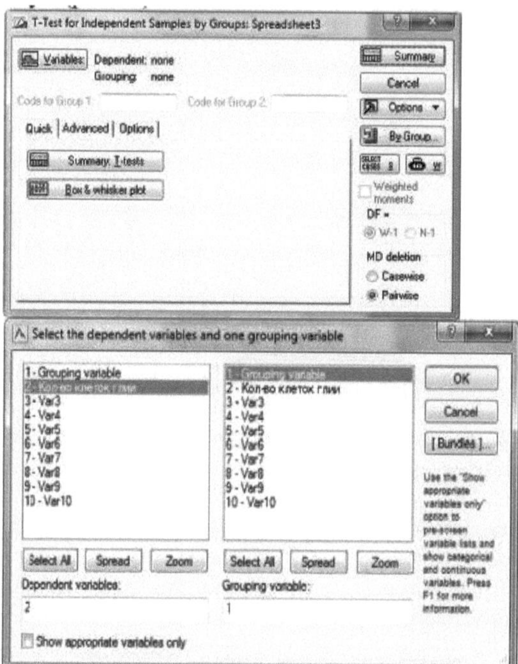

Figure 16. Dialogue windows for selecting the variables under study

3.   Click on the button **Summary: T-tests**. As a result, the programme will create a table containing the following results (Fig.17).

Variable	T-tests; Grouping: Grouping variable (Spreadsheet3) Group 1: 1 Group 2: 2										
	Mean 1	Mean 2	t-value	df	p	Valid N 1	Valid N 2	Std.Dev. 1	Std.Dev. 2	F-ratio Variances	p Variances
Кол-во клеток глии	41.40000	88.75000	-3.85696	11	0.002622	5	8	10.59717	25.70575	5.884111	0.106096

Fig. 17. Table with the results of t-test

This table contains the following indicators: Std. dev. standard deviation of sample 1; Std. dev. standard deviation of sample 2; P, Variances - probability of error for Fisher's F-test, if P > 0.05, the condition of homogeneity of variances is fulfilled.

The main indicator is the P value (probability to erroneously reject the null hypothesis of no differences between the averages). In our case, $P < 0.05$, therefore, there are statistically significant differences between the mean values of the number of glia cells in young and old animals.

Comparison of two "independent" groups whose data distribution does not follow a "normal" distribution (Mann-Whitney U- test).

If the distribution of the value of a trait in the two groups being compared differs from the "normal" distribution, the application of the parametric t-test for their comparison will lead to skewed results. In such cases, the appropriate non-parametric analogue of Student's t-test should be used.

Comparisons between two independent groups whose data distribution does not follow a "normal" distribution are made using the Mann- Whitney U-test.

The algorithm for selecting this method of statistical analysis can be presented in the form of the following scheme (Fig. 18):

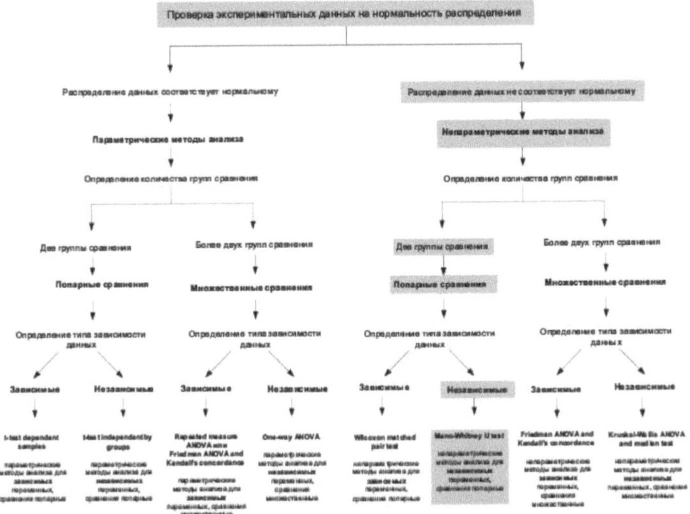

Fig. 18: Scheme for choosing the method of statistical analysis when comparing two independent groups whose data do not obey the law of normal distribution.

In the STATISTICA programme this test is performed as follows: enter the results of the study in the table according to the rules of data design for independent groups (see pages 13-14).

1. In the **Statistics** menu, select **Nonparametrics** (Figure 19).

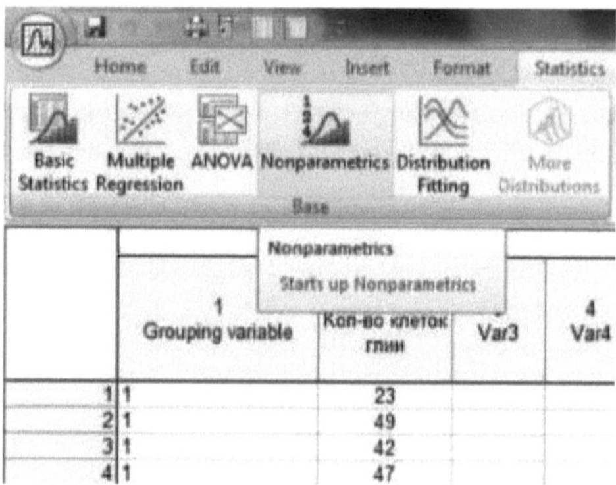

Fig. 19. Non-parametric statistics main menu section

2. Next, select Comparing **two independent samples** (Figure 20).

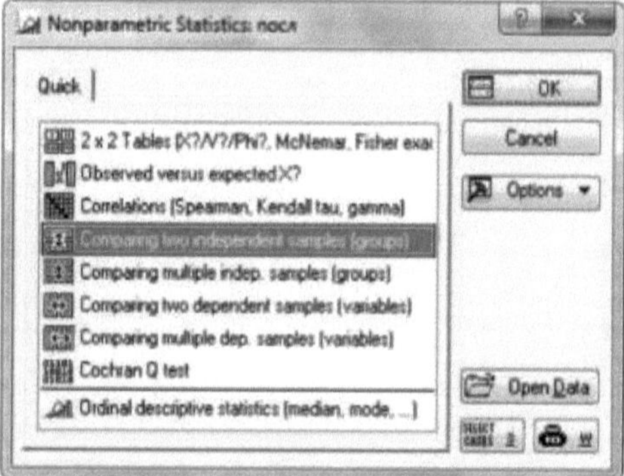

Fig. 20. The dialogue window of the Basic Statistics module

3. In the window that appears, click on the **Variables** button and select the dependent and group variables. In the right window select the group variable, in the left window select the dependent variable and click OK (Fig. 21)

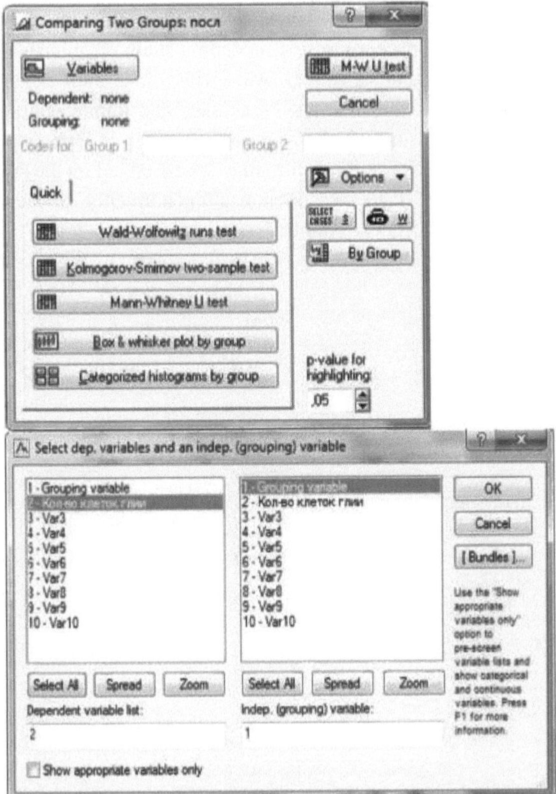

Fig. 21. Dialogue windows for selecting the variables to be tested 4. Click on the **Mann-Whitney U-test** or M-W U test button (Fig. 21), after which a table with the results will appear Fig. 22.

variable	Mann-Whitney U Test (nocn) By variable Grouping variable Marked tests are significant at p < .05000									
	Rank Sum Group 1	Rank Sum Group 2	U	Z	p-value	Z adjusted	p-value	Valid N Group 1	Valid N Group 2	2*1sided exact p
Кол-во клеток глии	19,00000	72,00000	4,000000	-2.26897	0.023271	-2.26897	0.023271	5	8	0.018648

Fig. 22. Table with t-test results

The main parameter to be considered is the probability of error (p-value). Since P < 0.05, there are statistically significant differences between the compared samples (Note: unlike the t-test, the Mann-Whitney test compares the sums of the ranks of each sample, rather than the mean values of the samples).

Comparison of the two "dependent" groups, the distribution of data in which corresponds to "normal" (T-test, dependent samples).

As a reminder, a researcher deals with dependent samples if the research is carried out on the same objects. Consider the following example.

It is known that intensive physical activity leads to weakened immunity and frequent colds. To find out the reasons for this phenomenon, a modelling study on animals was carried out. As a physical load, laboratory animals (mice) swam with a load until exhaustion. "Before" and "after" the load, blood was drawn from the animals and immunoglobulin levels were assessed. It is necessary to find out whether the mean number of immunoglobulins differs between "before" and "after" exercise. Since the study is carried out on the same animals, the samples are dependent. Provided that the requirements for "normality" of data distribution are met, we will use the t-test for dependent samples.

The algorithm for selecting this method of statistical analysis can be presented in the form of the following scheme:

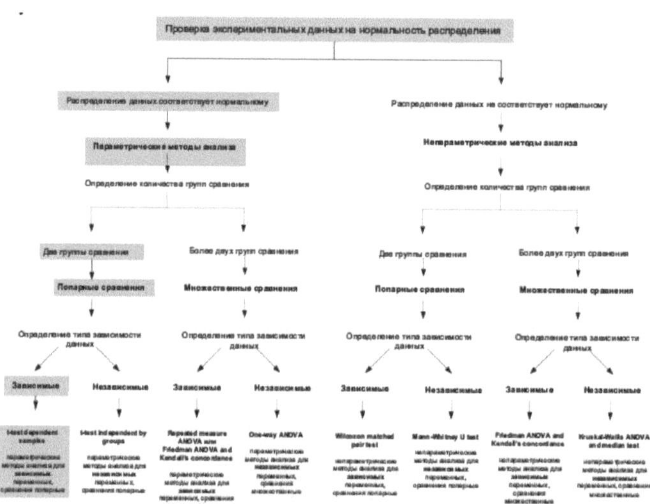

Fig. 23: Scheme for choosing the method of statistical analysis when comparing two dependent groups, provided that the data conform to the law of normal distribution
я

Note: Since the experimental data are **dependent,** enter the results of the study for each variable in separate columns (Figure 24). It is not necessary to include the group variable in the table. The data are similarly labelled in all cases where dependent groups are compared.

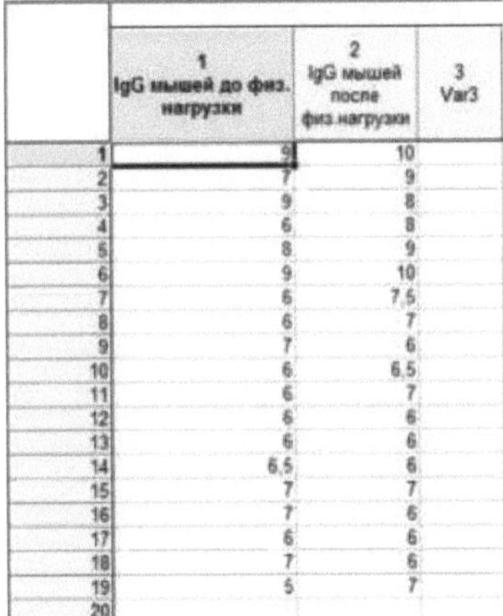

Figure 24 Example of data design when comparing two dependent groups.

To perform this variant of the t-test, you need to:
1. Start the **Basic statistics / t-test, dependent samples** module from the **Statistics** menu (Fig.
25).

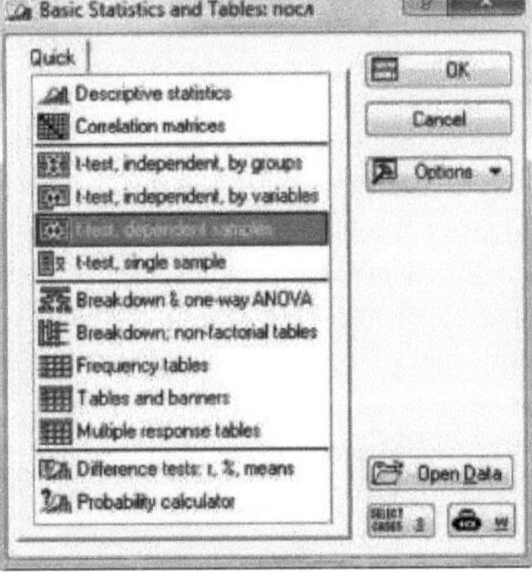

Fig. 25. The dialogue window of the Basic Statistics module

2. Click on the **Variables** button and specify the variables involved in the analysis: **First variable** and **Second variable** (Fig. 26).

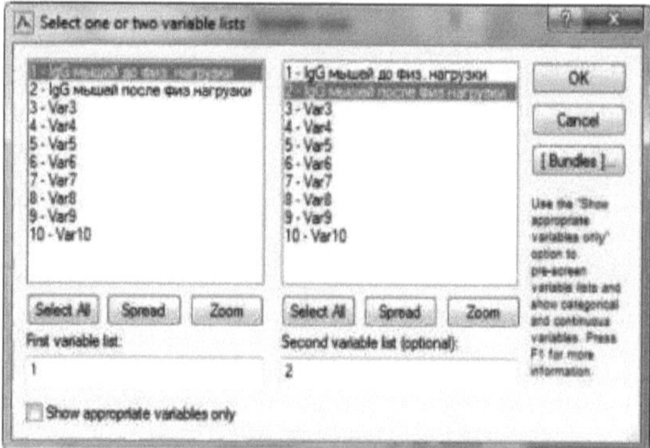

Fig. 26. Dialogue window for selecting the variables under study

3. click on the button **Summary: T-tests**. A table with the results will appear, similar to what we saw when performing the t-test for independent samples (Fig. 27).

| Variable | T-test for Dependent Samples (посл) Marked differences are significant at p < ,05000 | | | | | | | | | |
	Mean	Std.Dv.	N	Diff.	Std.Dv. Diff.	t	df	p	Confidence -95,000%	Confidence +95,000%
IgG мышей до физ. нагрузки	6,815789	1,169170								
IgG мышей после физ.нагрузки	7,263158	1,378087	19	-0,447368	1,052705	-1,85240	18	0,080439	-0,954756	0,060019

Fig. 27. Table with t-test results

Since in our case P > 0.05, we can conclude that the mean amount of immunoglobulins before and after physical activity is not significantly different.

Comparison of two dependent groups whose data distribution does not follow a "normal" distribution (Wilcoxon matched pair test).

If the distribution of data in two dependent samples differs from "normal", the Wilcoxon matched pair test should be used to compare them.

The algorithm for selecting this method of statistical analysis can be presented in the form of the following scheme:

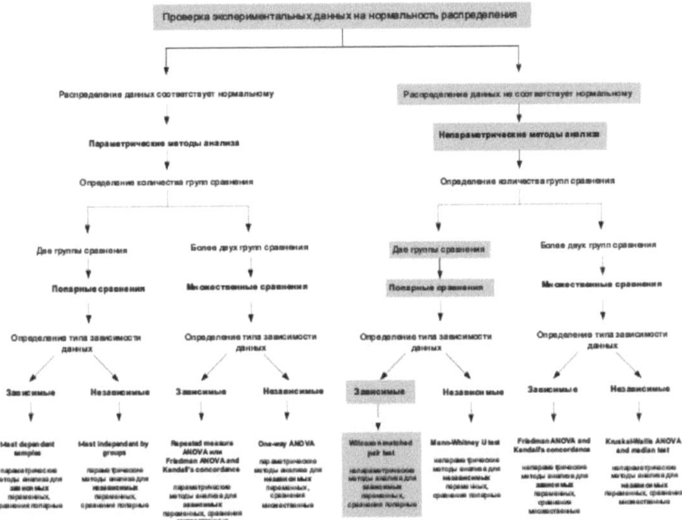

Fig. 28. Scheme for choosing the method of statistical analysis when comparing two dependent groups under the condition that the data do not conform to the law of normal distribution

To demonstrate the work of this test we will use the previous example, but let us assume that the condition of "normality" of the experimental data distribution is not fulfilled.

The Wilcoxon test can be run as follows: enter the results of the study in the table according to the rules of data formatting for dependent groups (see pages 19-20).

1. Under **Statistics / Nonparametrics / Comparing dependent samples** select t-test for dependent samples (Fig. 29).

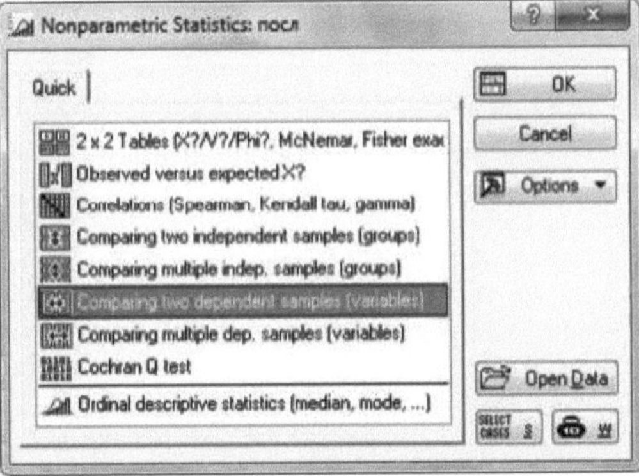

Fig. 29. The dialogue window of the Basic Statistics module 2. Click the **Variables** button, set the variables to be analysed and click the Wilcoxon **matched pair test** button (Fig. 30).

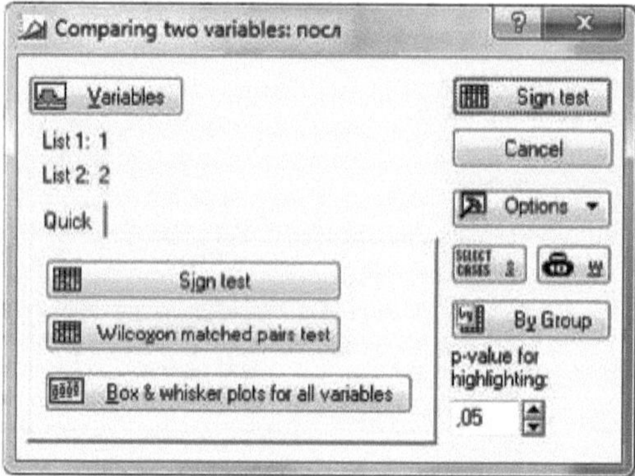

Fig. 30. The dialogue window for selecting the variables under study The table will appear as a result (Fig. 31). Since P>0.05, therefore there are no statistically significant differences between the compared samples.

Pair of Variables	Wilcoxon Matched Pairs Test (посл) Marked tests are significant at p <,05000			
	Valid N	T	Z	p-value
IgG мышей до физ. нагрузки & IgG мышей после физ.нагрузки	15	29,50000	1,732284	0,083224

Figure 31: Table with the results of the Wilcoxon test.

CHAPTER 3

Multiple comparisons (comparisons of several groups).

The Student's t-test and its non-parametric analogues discussed above are designed to compare **only two samples**. However, very often this test is used for comparisons in 3 or more samples, which sharply increases the probability of the first kind of error (the first kind of error is the probability of falsely rejecting the null hypothesis, i.e. finding differences where there are none). The maximum permissible probability of this error is 5%.

Let's assume that it is necessary to make comparisons of 3 independent groups. For this purpose it is supposed to carry out 3 pairwise comparisons: gr. 1 x 2; gr. 1 x 3 and gr. 2 x 3. This means that the control of the first kind of error can be ensured only by dividing the value of the nominal significance level by the number of pairwise comparisons. In this case 3, we get 0.05/3=0.017. Thus, the null hypothesis is rejected if the achieved level of significance using paired Student's criterion P< 0.017.

To avoid this error it is necessary to use special methods of statistical analysis for multiple comparisons. The algorithm of choosing this method of statistical analysis can be presented in the form of the following scheme:

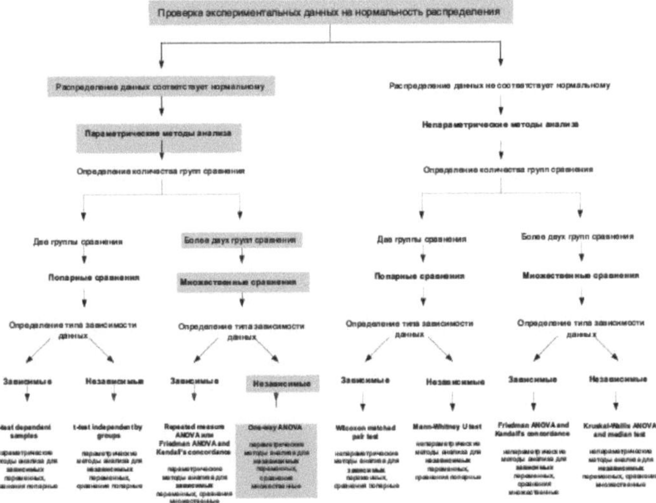

Fig. 32. Scheme for choosing the method of statistical analysis when comparing three or more independent groups, provided that the data conform to the law of normal distribution

One-factor analysis of variance (One-way ANOVA).

As an example, requiring the use of analysis of variance, we can consider the study of blood biochemical parameters in several groups of patients suffering from Parkinson's disease, as well as in a group of conditionally healthy individuals. The first group - conditionally healthy persons (control), the second group - patients at the early stage of the disease, the third group - patients at the late stage of the disease. Such biochemical index as the amount of alpha-synuclein in blood plasma was investigated in these groups.

Since there are more than two study groups and, consequently, more than two comparison groups, we will use a one-factor analysis of variance. To perform this type of analysis, the results of the study must be entered into the table in accordance with the rules of data processing for independent groups (see page 14).

1. From the **Statistics** menu, start the **One-way ANOVA** module (Figure 33). Next, select the type of analysis **One-way ANOVA**.

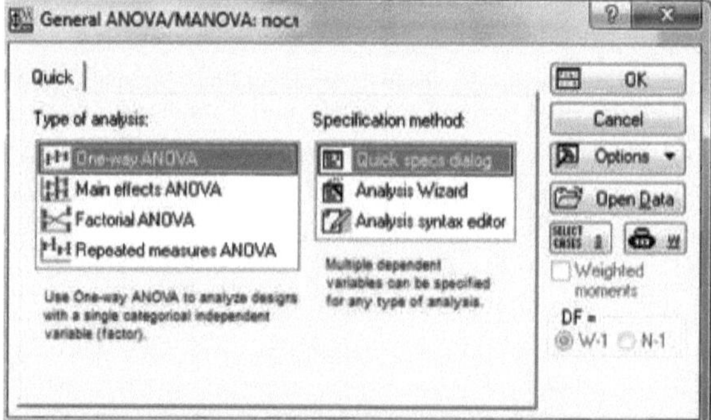

Fig. 33. The variance analysis module dialogue window 2. Click on the **Variables** button and select the dependent and grouping variables

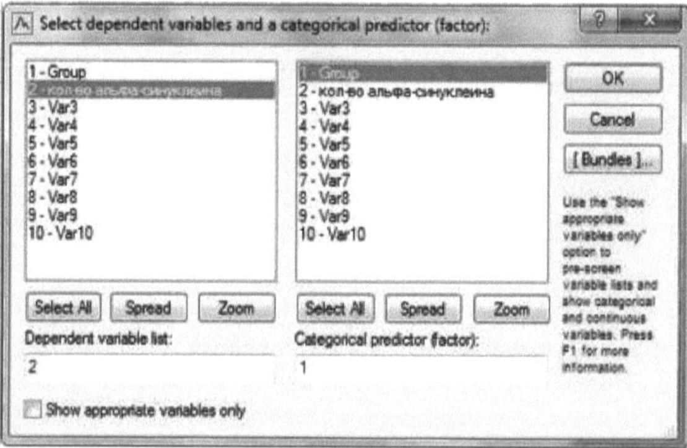

Figure 34. Dialogue window for selecting the variables to be analysed 3. Click on the buttons: **Factor codes** / **All** (this will indicate to the software that all experimental groups should be analysed) / **OK** / **OK** (Fig. 35).

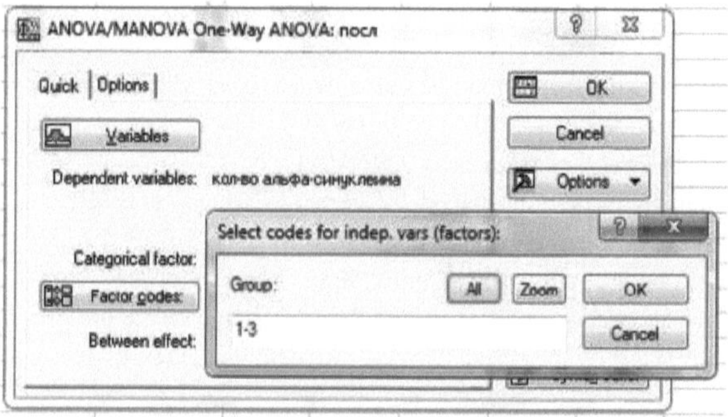

Figure 35. Dialogue window for selecting codes of the studied groups

As a result, a window with 8 tabs will appear (Fig. 36), automatically opened on the **Quick** tab.

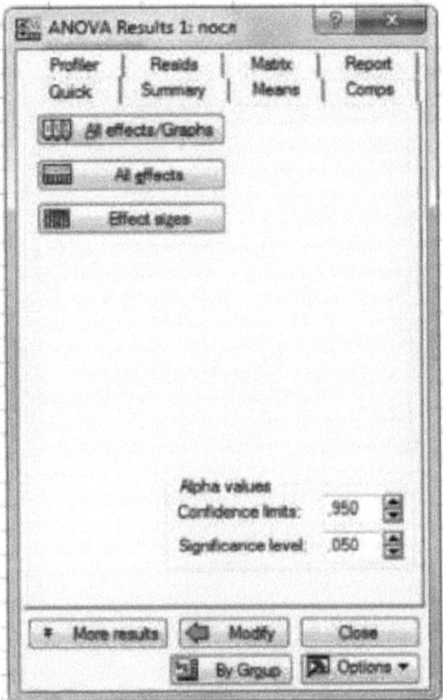

Figure 36. The dialogue window for selecting the results of variance analysis

By clicking on the **All effects** button you can quickly get the results of the analysis. However, the analyses under consideration are parametric and therefore require a number of prerequisites to be fulfilled:

1) Homogeneity of variance (no statistically significant difference between the scatter measures of the data within groups);

2) Subjection of the data (in all groups) to the law of normal distribution;

3) The independent nature of the samples.

Therefore, the sample should be checked to ensure that it meets these requirements. To check for homogeneity of variance, it is necessary to:

1. Click on the **More results** button located at the bottom of the **ANOVA Results** window.

2. In the window that appears (Fig. 37), open the Assumptions tab.

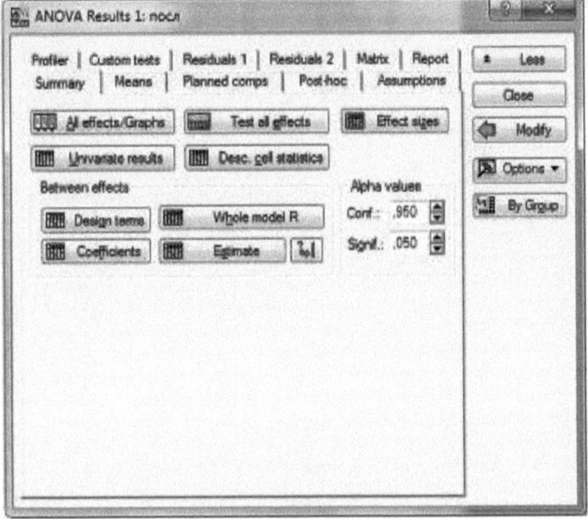

Figure 37. Additional variance analysis results dialogue window

3. In the **Homogeneity of variances/covariances** section, click on **Levene's test** (Fig. 38).

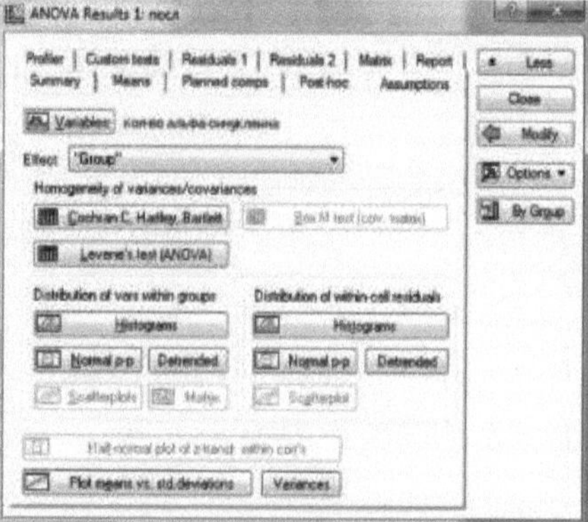

Figure 38. Additional variance analysis results dialogue window

The table that appears shows the results of the variance comparison test (Fig. 39). If there are no differences between dispersions (P>0.05), then the use of parametric variant of variance analysis is justified. In our case, there are no differences (P = 0.12).

	Levene's Test for Homogeneity of Variances (посл) Effect: "Group" Degrees of freedom for all F's: 2, 21			
	MS Effect	MS Error	F	p
кол-во альфа-синуклеина	35,19565	15,09321	2,331887	0,121741

Figure 39. Results of the test for differences between dispersions

To check the "normality" **of** the distribution of the analysed data, you can use the option available in the **Distribution of variables within groups** field. However, it is better to perform this operation in advance, using a special module - **Distribution fitting**.
If all conditions are met, click the **Test all effects** button on the **Summary** tab (Fig. 40).

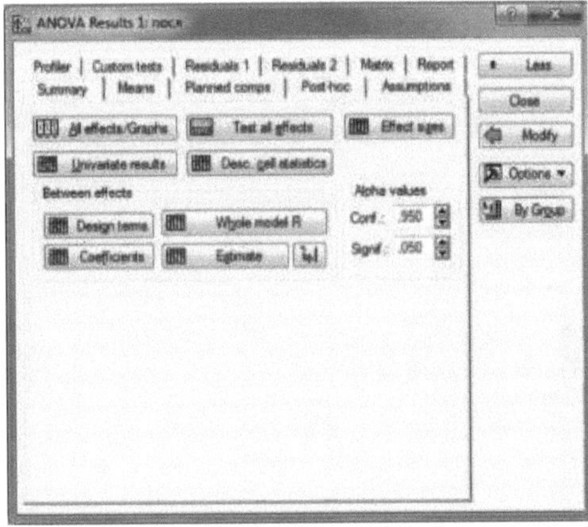

Figure 40. The dialogue window for selecting the results of variance analysis

In the appeared table (Fig. 41) it is necessary to find the cell with the error value P. Since in our example P < 0.05, we can conclude that the amount of alpha-synuclein in the plasma of patients of different groups is statistically significantly different.

Effect	Univariate Tests of Significance for кол-во альфа-синуклеина (Sigma-restricted parameterization Effective hypothesis decomposition				
	SS	Degr. of Freedom	MS	F	p
Intercept	1834.555	1	1834.555	44.44760	0.000001
Group	498.408	2	249.204	6.03772	0.008482
Error	866.766	21	41.275		

Figure 41. Results of analysis of variance

Post-hoc analysis (Post-hoc analysis).
When conducting analysis of variance, it is important to realise that it can only test the hypothesis that there are no differences between the groups being compared as a whole. It is not possible to find out which groups differ from each other. To find out this question, multiple comparison methods are used, which are part of the so-called posterior analysis **Post-hoc analysis**. These methods allow pairwise comparisons of the mean values of all groups included in the analysis of variance.

To perform posterior comparisons in the **ANOVA Results** window, click the **More results** button (Fig. 36). Next, open the **Post hoc** tab (**posterior** comparisons) (Fig. 42).

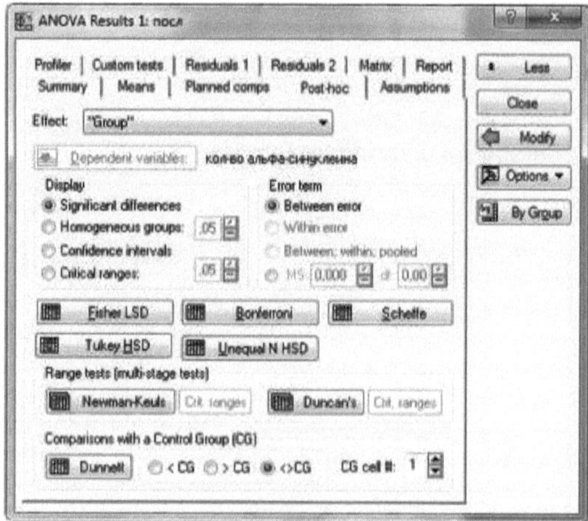

Figure 42. The dialogue window for selecting the methods of posterior analysis

The STATISTICA programme offers several varieties of multiple comparison tests: **Fisher LSD, Bonferroni, Scheffe, Tukey HSD, Newman-Keuls, Duncan's, Dunnet** (they differ somewhat in power). The **Tukey HSD** and **Newman-Keuls** tests are the most commonly used.

By clicking on the button of the corresponding test, you can get a table with a matrix of P values (Fig. 43).

Cell No.	Group	{1} 15.159	{2} 5.0133	{3} 6.0562
	LSD test: variable кол-во альфа-синуклеина (посл) Probabilities for Post Hoc Tests Error: Between MS = 41.275, df = 21.000			
1	1		0.004737	0.009943
2	2	0.004737		0.748631
3	3	0.009943	0.748631	

Figure 43. Results of the posterior analysis

Figure 43 shows that a statistically significant difference in the amount of alpha-synuclein is observed between the control group and patients at the first stage of the disease, as well as between the control group and patients at the second stage of the disease. There is no statistically significant difference in the amount of alpha-synuclein between the groups of patients. Friedman ANOVA and Kendall's concordance analysis of variance (Friedman ANOVA and Kendall's concordance).

Friedman ANOVA is used if the samples under study are interrelated (dependent). It is important to note that, being non-parametric, it does not require compliance with the conditions of "normality" of distribution and homogeneity of dispersions in the studied groups.

The algorithm for selecting this method of statistical analysis can be presented in the form of the following scheme:

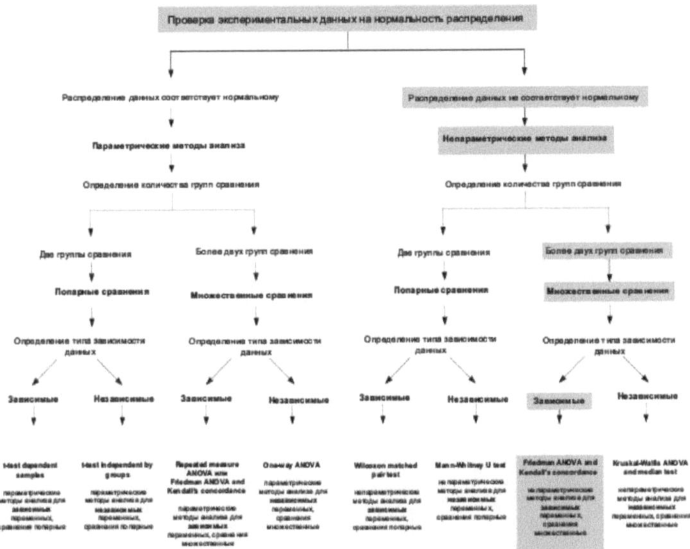

Fig. 44. Scheme for choosing the method of statistical analysis when comparing three or more dependent groups under the condition that the data do not conform to the law of normal distribution

Fig. 44 presents data on changes in the amount of blood phospholipids in athletes during the training process, as well as in the competitive period 20112012. It is necessary to find out whether there were significant changes in the amount of phospholipids by the end of the sports season. Since the same athletes participated in the study, the obtained samples are interrelated (dependent). We use Friedman's analysis of variance to compare them.

Let us enter the results of the study into the table in accordance with the rules of data formatting for dependent groups (see pages 20-21) (Fig. 45).

	1 24.11.2011	2 25.12.2011	3 22.01.2012	4 23.02.2012
1	1,19	1,77	1,56	1,22
2	1,8	1,09	1,39	2,21
3	1,9	1,21	1,34	2,89
4	2,12	1,02	2,24	2,3
5	1,06	0,72	1,7	2,1
6	1,5	1,3	1,47	1,11
7	1,7	1,4	1,6	1,4
8				

Figure 45. Example of data layout when comparing three or more dependent groups

1. Start the analysis module from the menu : **Statistics / Nonparametric / Comparing multiple dependent samples** (Figure 46).

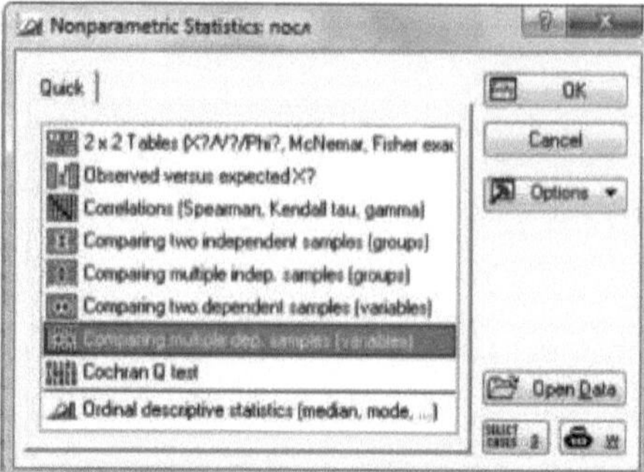

Figure 46. Dialogue window of the variance analysis module

2. Click the **Variables** button and select the variables to be analysed (Fig. 47).

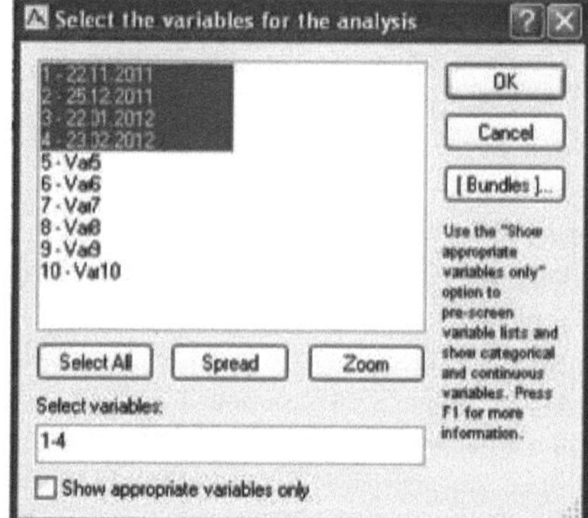

Figure 47. Dialogue window for selecting the variables under study

3. Click on **Summary: Friedman** ANOVA **and Kendall's concordance** (Figure 48).

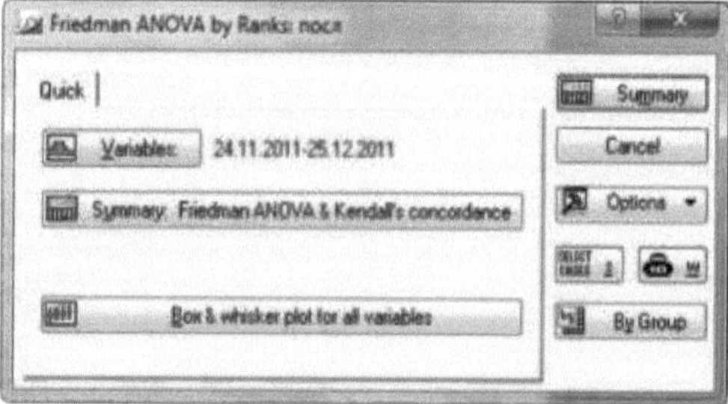

Figure 48. Dialogue window for launching the analysis of variance

2.   In the table with the results, find the error value P (located in the table header) (Fig. 49). Since in our case P>0.05, therefore, there are no reliable differences between the amount of phospholipids in the studied groups.

In the same heading, the so-called Kendall's Coeff. of Concordance (Coeff. of Concordance) is given. The closer the Kendall coefficient is to 1, the more

30

differences between the groups.

	Friedman ANOVA and Kendall Coeff. of Concordance (Spreadsheet1) ANOVA Chi Sqr. (N = 7, df = 3) = 4,304348 p = ,23042 Coeff. of Concordance = ,20497 Aver. rank r = ,07246						
Variable	Average Rank	Sum of Ranks	Mean	Std.Dev.			
22.11.2011	2,714286	19,00000	1,610000	0,382840			
25.12.2011	1,642857	11,50000	1,215714	0,328677			
22.01.2012	2,714286	19,00000	1,614286	0,302316			
23.02.2012	2,928571	20,50000	1,890000	0,659798			

Figure 49. Results of analysis of variance

However, this method has the same disadvantage as its parametric analogue (one-factor analysis of variance). It allows testing only the hypothesis that there are no differences between the compared groups as a whole. This type of analysis does not allow us to find out which groups differ from each other.

Kruskal-Wallis analysis of variance (Kruskal-Wallis ANOVA).

As noted above, experimental data obtained in biological research rarely obey the law of normal distribution. Moreover, very often the sample size is too small to draw any conclusions regarding the type of distribution. All this makes the application of parametric analysis of variance impossible.

One way out of this situation is to apply Kruskal-Wallis nonparametric analysis of variance (or H-test) (Kruskal- Wallis ANOVA).

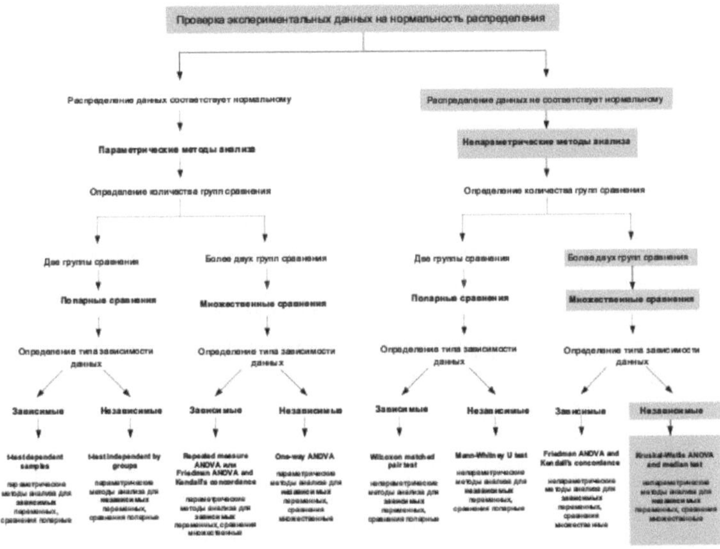

Fig. 50. Schematic of the choice of statistical analysis method when comparing three or more independent groups, provided the data do not conform to the law of normal distribution Let us examine the application of Kruskal-Wallis analysis of variance on the following example: Fig. 51 shows data on the number of glial cells in different brain structures. 51 shows data on the number of glial cells in different brain structures.

	1 Group	2 кол-во глиальных клеток
1	черн. субстанц.	87
2	черн. субстанц.	109
3	черн. субстанц.	41
4	черн. субстанц.	79
5	черн. субстанц.	106
6	черн. субстанц.	126
7	черн. субстанц.	85
8	черн. субстанц.	77
9	красн. ядра	32
10	красн. ядра	15
11	красн. ядра	20
12	красн. ядра	18
13	красн. ядра	24
14	красн. ядра	25
15	интраламинар. ядра	27
16	интраламинар. ядра	38
17	интраламинар. ядра	27
18	интраламинар. ядра	29
19	интраламинар. ядра	51
20		

Figure 51. Example of data layout when comparing three or more independent groups

As can be seen, the number of observations in each department is small (n not more than 8), which does not allow us to correctly assess the nature of data distribution. If we resort to a small trick and combine all the available data into a single population, it turns out that their distribution does not follow the law of normal distribution.

To find out whether the number of cells differs in different parts of the brain, we apply Kruskal-Wallis analysis of variance. Note that the data are entered in the table according to the rules of design for independent groups (see pages 13-14).

1. Start the **Statistics / Nonparametrics / Comparing multiple independent samples** module from the menu (Figure 52).

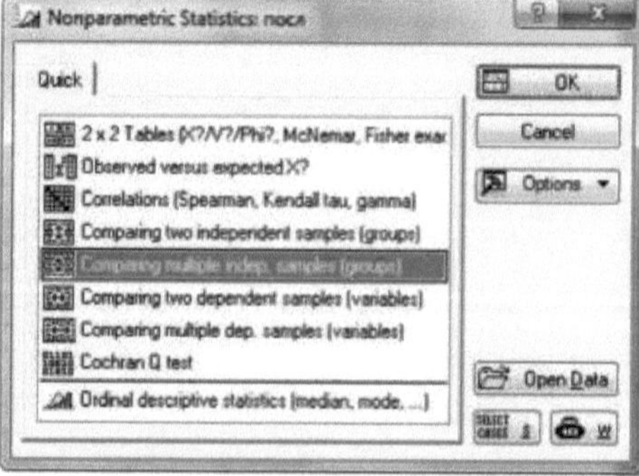

Figure 52. Dialogue window for selecting analysis of variance

2. Click the **Variables** button and select the dependent ("number of cells") and grouping variables (Figure 53).

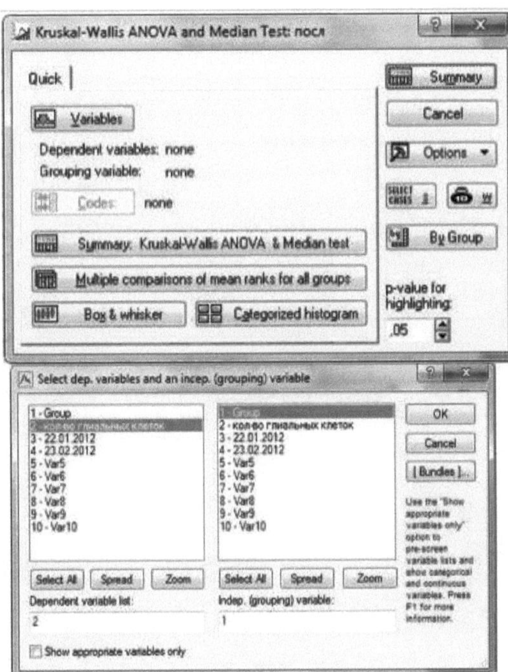

Figure 53. Dialogue windows for selecting the studied variables 3. Then click on the buttons: **Factor codes / All / OK** (Fig. 54).

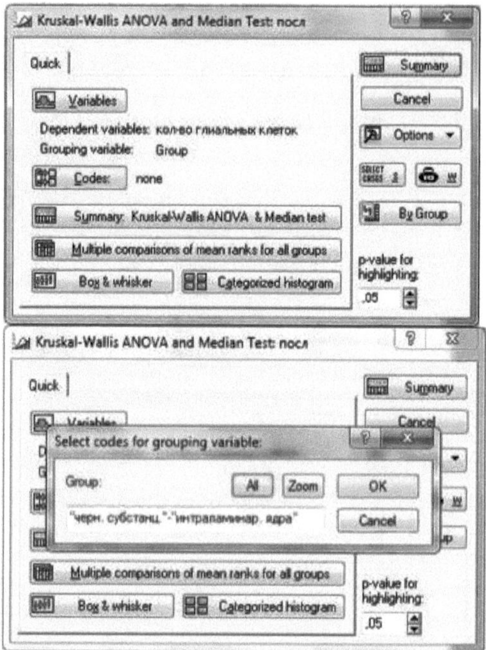

Figure 54. Dialogue windows for selecting codes of the studied groups

4. Click on **Summary**: Kruskal-Wallis ANOVA **and Median test** (Figure 55).

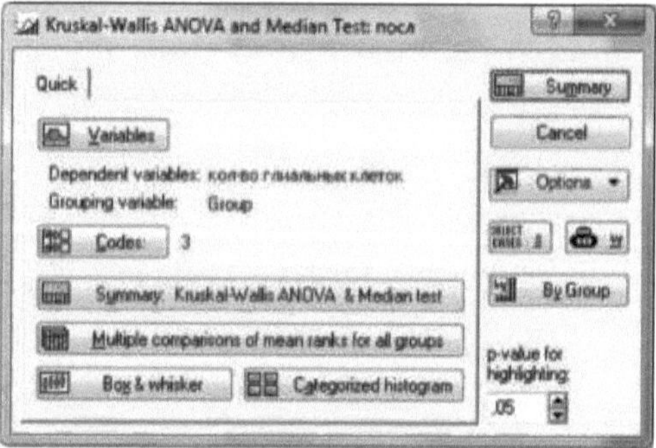

Figure 55. Dialogue window for launching the analysis of variance

4. In the table with the results (Fig. 56), find the error P value for the null hypothesis that the number of glial cells in different brain structures does not differ.

Depend.: кол-во глиальных клеток	Kruskal-Wallis ANOVA by Ranks; кол-во глиальных клеток (посл) Independent (grouping) variable: Group Kruskal-Wallis test: H (2, N= 19) =14,44188 p = ,0007			
	Code	Valid N	Sum of Ranks	Mean Rank
черн. субстанц.	101	8	123,0000	15,37500
красн. ядра	102	6	24,0000	4,00000
интраламинар. ядра	103	5	43,0000	8,60000

Is. 56. Results of analysis of variance

If $P<0.05$ (as in our example), the studied groups are statistically significantly different from each other. In addition to the results of the Kruskal-Wallis test, the programme offers the results of the so-called median test. It tests the same null hypothesis, but is less powerful.

Factorial ANOVA (Factorial ANOVA).

Factor analysis is designed to identify the influence of various factors (conditions) or their combination on the change of the studied attribute. In other words, it allows to study the dependence of any quantitative (dependent) attribute on one or more qualitative attributes (factors).

Let's look at how factor analysis works, using the following example: let's say we need to answer the question: How do age, gender and education level affect blood pressure?

Suppose that all subjects who participated in the study can be divided into 4 groups according to their age:

gr. 1: up to 30 years;

gr. 2: 31 to 40 years old;

gr. 3: from 41 to 50 years old;

gr. 4: more than 51 years.

According to the level of education, the subjects can be divided into 3 groups: gr. 1: higher education;

gr. 2: secondary education;

gr. 3: without education.

In addition, we know the systolic blood pressure (SBP) of each study participant as well as their gender.

Thus, ADS is a   quantitative trait, while "Age", "Gender" and "Education" are the factors whose influence we need to investigate.

Before proceeding to the analysis, let's enter the data into the table. Note that all data are arranged in vertical columns. The first 3 columns are qualitative attributes (factors) or grouping variables, the fourth column is the dependent variable (Figure 57).

	1 пол	2 возраст	3 образование	4 САД
1	M	1	c	120
2	W	1	v	125
3	M	1	n	130
4	M	2	c	141
5	M	3	v	160
6	M	4	n	161
7	M	4	n	135
8	W	3	c	123
9	M	1	v	130
10	M	2	v	128
11	M	2	v	141
12	M	3	c	161
13	M	3	n	167
14	W	4	n	125
15	W	1	n	129
16	W	2	v	142
17	M	3	c	165
18	M	4	c	161
19	M	4	c	159

Figure 57. Example of data design for factor analysis

1. From the **Statistics / ANOVA** menu, start the **Factorial ANOVA** module (Figure 58).

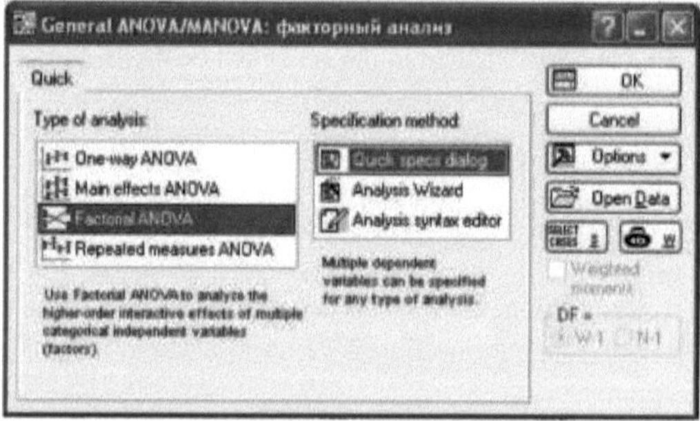

Figure 58. Dialogue window for selecting factor analysis

2. In the window that appears, click on the **Variables** button and select the grouping variables (Gender, Age, Education) and the dependent variable (ADS) (Figure 59).

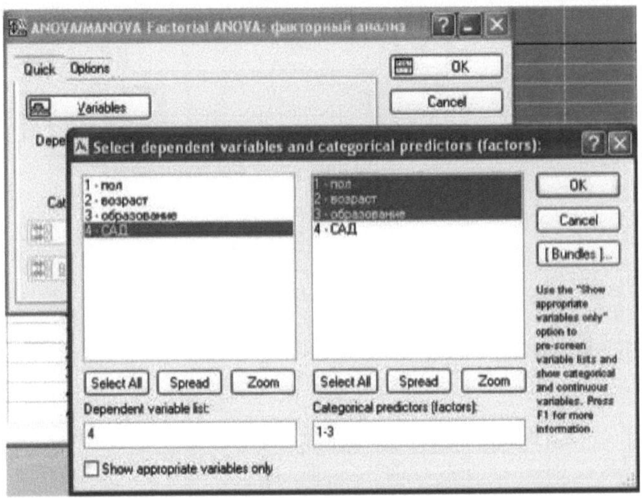

Figure 59. Dialogue window for selecting grouping and dependent variables

3. Next, click the **Factor code** button, specify the codes of the factors, the influence of which should be evaluated. Since there are not very many factors, we click the **All button** (Fig. 60).

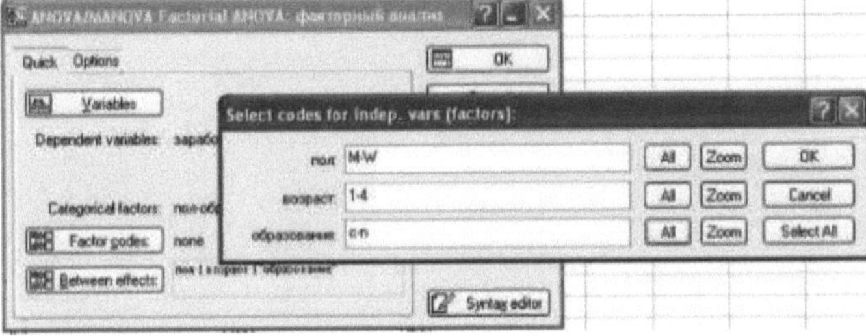

Figure 60. Dialogue window for selection of factor codes

4. You don't have to set the factor codes: if you click **OK**, the programme will set them automatically. As a result, a window with 8 tabs will appear. After selecting the **Summary tab,** click the **Test All effects** button (Fig. 61).

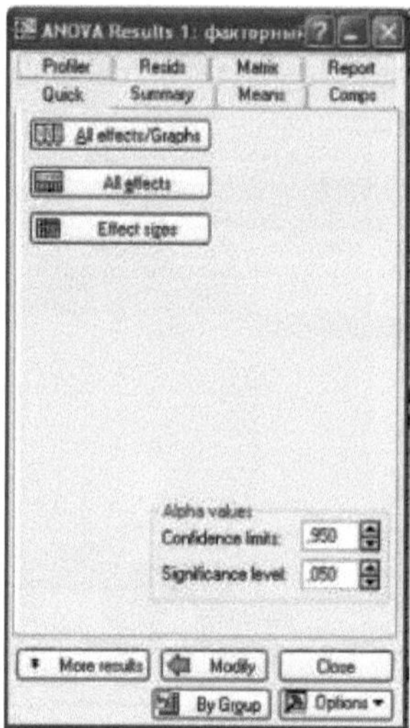

Figure 61. Dialogue window for selection of factor codes

As a result, a table with the results of analysis of variance will appear (Fig. 62). At the end of the rows of this table, the error probabilities for the null hypotheses about the absence of influence of the facts "Gender", "Age" and "Education" on the level of CAD are given.

Effect	Univariate Tests of Significance for САД (факторн Sigma-restricted parameterization Effective hypothesis decomposition				
	SS	Degr. of Freedom	MS	F	p
Intercept	1267884	1	1267884	8587.862	0.00000
пол	18	1	18	0.123	0.72720
возраст	1258	3	419	2.839	0.04356
образование	283	2	141	0.957	0.38852
пол*возраст	1606	3	535	3.626	0.01676
пол*образование	369	2	185	1.250	0.29237
возраст*образование	671	6	112	0.758	0.60526
пол*возраст*образование	1555	6	259	1.755	0.12006
Error	11073	75	148		

Figure 62. Results of factor analysis

The table shows that "Age" has the most significant effect on CAD: P<0.05. We failed to prove a similar effect for "Gender" and "Education" in this experiment (P>0.05). The rows "Gender\ Age" etc. concern the mutual influence of the studied factors on CAD. As can be seen, the greatest interaction (mutual influence) on the value of CAD is exerted by the factors "Gender" and "Age" ($P > 0.05$).

To assess differences in mean CAD values across categories, graphical means can be used.

1. Click on the **All effects/Graphs** button. The resulting window (Figure 63) lists all the effects under consideration. Statistically significant effects are marked with *.

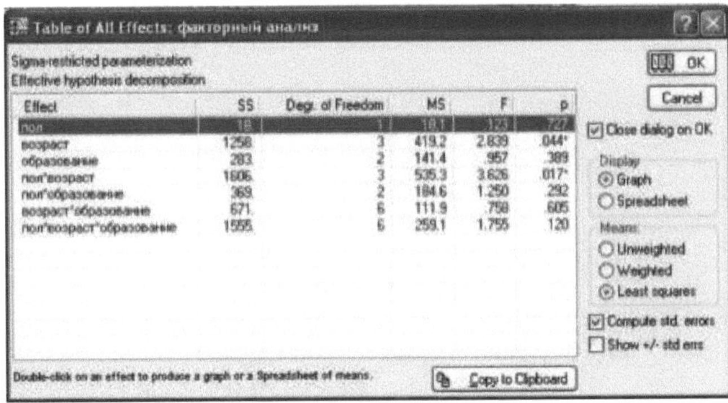

Figure 63. Additional results of factor analysis

2. Select the **Age** effect, in the **Display** group, specify **Spreadsheet**, and click **OK.**
The table that appears shows for each effect level the mean values of the dependent variable CAD, the standard error and the confidence limits (Fig. 64).

Cell No.	age; LS Means (factor analysis) Current effect: F(3, 75)=2.8392, p=.04358 Effective hypothesis decomposition				
	age	SAD Mean.	SAD Std. Err.	GARDEN -95 00%	SAD +95 00%
1	1	135.6667	3.788611	128.1194	143.2140 15
2	2	141.6071	3.061660	135.5080	147.7063 25
3	3	145.3899	3.033626	139.3466	151.4332 28
4	4	147.7734	2.227169	143.3367	152.2102 31

Figure 64. Table of systolic blood pressure (SBP) in all age groups
It is convenient to present this table in a graphical form. To do this, select **Graph** in the **Display** group and click **OK.** The corresponding graph will appear (Figure 65).

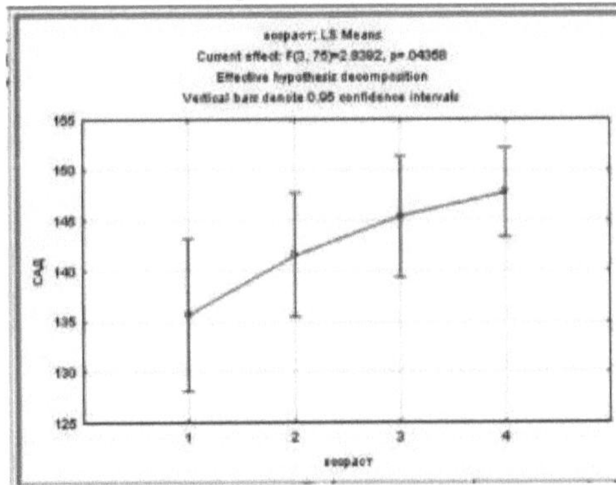

Figure 65. Graph of average income indicators in different age groups
The graph clearly shows how the mean values of CAD change in different age groups (depending on the "Age" factor).
In addition to assessing the influence of the factors under study ('Age', 'Education' and 'Gender') and their interactions on SAD scores, it is important to understand what proportion of the variability they explain.
1. In the **Summary** tab, click on the Whole model R button (Figure 66) and the corresponding table will appear (Figure 67).

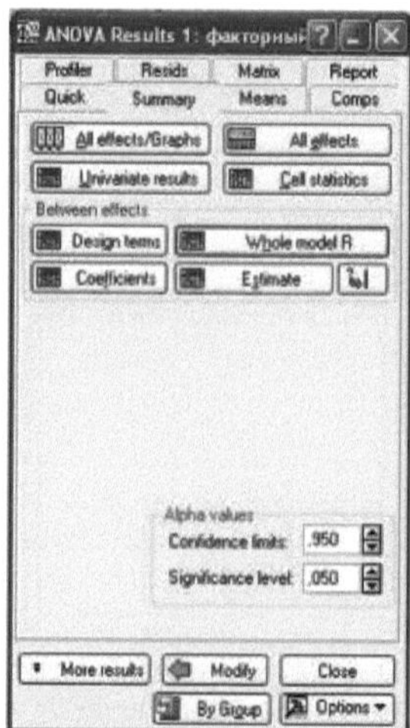

Figure 66. Dialogue window for selection of factor codes

Dependent Variable	Test of SS Whole Model vs. SS Residual (факторный анализ)										
	Multiple R	Multiple R?	Adjusted R?	SS Model	df Model	MS Model	SS Residual	df Residual	MS Residual	F	p
САД	0.586255	0.343695	0.142428	5798.595	23	252.1128	11072.76	75	147.6368	1.707656	0.043807

Figure 67. Table of SS model and SS residuals

The coefficient **R (Multiple R)**, the square of the multiple correlation coefficient or the coefficient of determination, is of most interest. It shows how much of the variability is explained by the model. The closer R is to one, the better the model is built. In our case R2 = 0.58, which indicates that the quality of the model is not very good. We can say that the level of CAD is 58% determined by the factors included in our model ("Gender", "Age" and "Education"). Nevertheless, other factors not included in our model also influence the level of CAD.

Analysis of variance with repeated measures ANOVA (Repeated measure ANOVA).

The type of analysis of variance described above applies only when there is only one dependent variable. If there are several dependent variables and they are the result of repeated measurements of the same trait, repeated measures analysis of variance methods are used.

Let us analyse this type of analysis using a similar example (see factor analysis). Let us evaluate the influence of the factors "Age", "Gender" and "Education" on the value of systolic blood pressure (SBP). Let's assume that CAD was measured twice after a certain period of time. Thus, we will deal with repeated measurement of the same trait.

Let's enter the data of repeated measurements in the table in a separate column (Fig. 68).

	1 пол	2 возраст	3 образование	4 САД	5 САД1
1	M	1	c	120	123
2	W	1	v	125	125
3	M	1	n	130	134
4	M	2	c	141	140
5	M	3	v	160	163
6	M	4	n	161	162
7	M	4	n	135	136
8	W	3	c	123	127
9	M	1	v	130	132
10	M	2	v	128	127
11	M	2	v	141	140
12	M	3	c	161	167
13	M	3	n	167	167
14	W	4	n	125	128
15	W	1	n	129	134
16	W	2	v	142	149
17	M	3	c	165	164
18	M	4	c	161	163
19	M	4	c	159	161
20	M	3	v	133	131
21	M	2	v	121	124
22	M	1	n	137	138
23	W	4	v	147	150
24	W	3	n	131	130
25	W	2	n	135	136

Figure 68. Example of data design for factor analysis with repeated measurements

1. From the **Statistics / ANOVA** menu, start the **Repeated measure ANOVA** module (Figure 69).

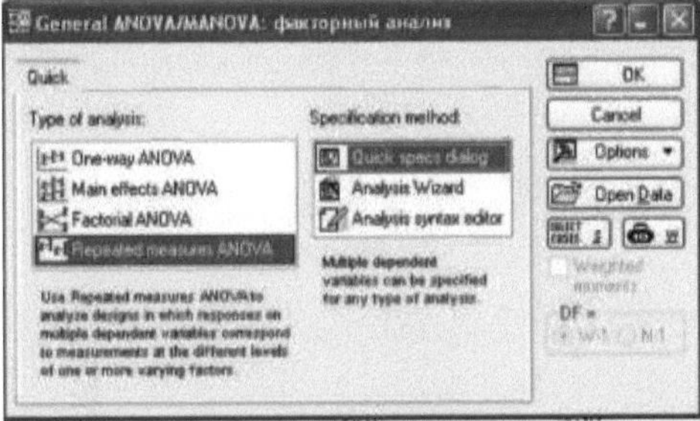

Figure 69. Dialogue window for selecting factor analysis with repeated measurements

2. In the window that appears, click the **Variables** button and select grouping variables ("Gender", "Age", "Education") and dependent variables (CAD and CAD1) (Fig. 70).

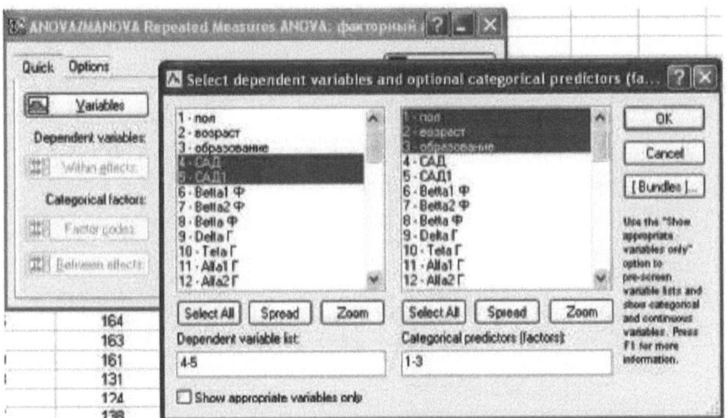

Figure 70. Dialogue window for selecting grouping and dependent variables 3. Click the **Within effects** button. In the opened window in the **Factor Name** field specify the name of

the repeated measurements factor, for example "Education" (by default the programme will offer to select one factor for repeated measurements with the name R). In the **No. of levels** field you can specify the number of repeated measurements, in our case there are 2 (Fig. 71).

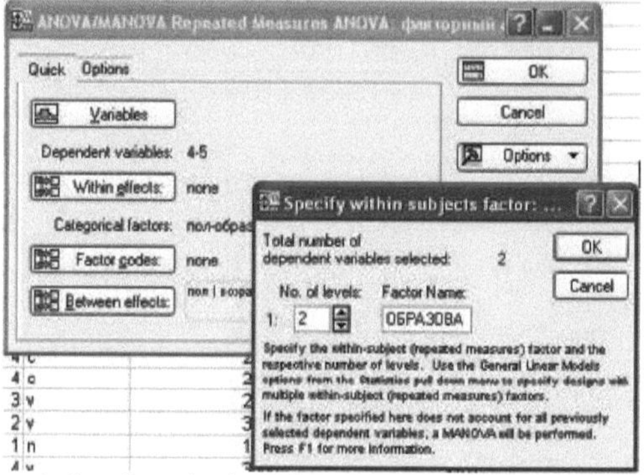

Figure 71. Dialogue window for selection of repeated measurements factor

3.   Press the **Factor code** button, specify the codes of the factors, the influence of which should be evaluated (Fig. 72). Since there are not very many factors, click the **All button**. You do not have to specify the factor codes: if you click **OK**, the programme will set them automatically.

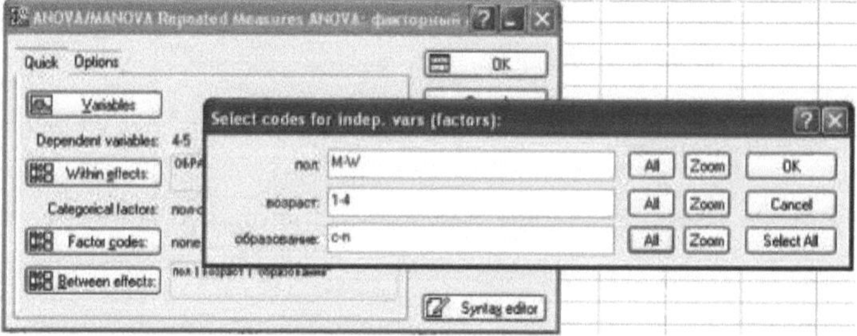

Figure 72. Dialogue window for selection of factor codes

4.   Press **OK**. This will bring up the familiar window with 8 tabs. Selecting the **Summary tab**, click the **Test All** effects button. *The* table shows that the hypothesis of inequality of mean is true for the factor "Age" and the combination of factors "Gender / Age". Such factors as "Education" and "Gender" separately have no effect on the value of CAD (Fig. 73).

Effect	Repeated Measures Analysis of Variance (факторн Sigma-restricted parameterization Effective hypothesis decomposition				
	SS	Degr. of Freedom	MS	F	p
Intercept	2574179	·	2574179	9215.447	0.000000
(1)пол	11	·	11	0.040	0.841285
(2)возраст	2471	3	824	2.949	0.038139
(3)образование	552	2	276	0.989	0.376776
пол*возраст	3005	3	1002	3.586	0.017583
пол*образование	642	2	321	1.160	0.322268
возраст*образование	1065	6	177	0.635	0.701564
пол*возраст*образование	2953	6	492	1.762	0.118506
Error	20950	75	279		
(4)ОБРАЗОВА	144	·	144	38.502	0.000000
ОБРАЗОВА*пол	7	·	7	1.885	0.173884
ОБРАЗОВА*возраст	2	3	1	0.152	0.927896
ОБРАЗОВА*образование	4	2	2	0.506	0.604664
ОБРАЗОВА*пол*возраст	10	3	3	0.852	0.469963
ОБРАЗОВА*пол*образование	11	2	5	1.405	0.251860
ОБРАЗОВА*возраст*образование	22	6	4	0.985	0.441355
4*1*2*3	7	6	1	0.329	0.919601
Error	281	75	4		

Figure 73. Results of factor analysis

To visualise the results obtained, as in the case of factor analysis, you can use various graphical effects by clicking on the **All effects/Graphs** button (Fig. 61).

CHAPTER 4

Correlation analysis.

In scientific research, it is often necessary to search for relationships between different attributes of the groups under study (rainfall and crop yield, height and weight of a person, body temperature and pulse rate, etc.). In the given examples, the attributes are related to each other, a change in one variable leads to a change in another.

To solve problems of this type, different varieties of correlation analysis are used. Correlation analysis allows to estimate the direction of the relationship between two features (direct or inverse), as well as to express it quantitatively using the correlation coefficient. The closer the coefficient is to 1 (modulo), the stronger the relationship between the attributes. The sign of the coefficient (+ or -) indicates the direction of dependence.

Pearson's correlation coefficient.

The Pearson correlation coefficient belongs to the group of parametric methods of statistical analysis and requires the following mandatory conditions:

1. The distribution of the data must obey the law of normal distribution;

2. The relationship between traits should be linear in nature.

You can check the data for "normality" of **the** distribution using the **Distribution fitting** module. This type of analysis is discussed in detail above. To assess the linearity of dependence between the attributes, you can use the Scatterplots module. This procedure is described in detail below in the Regression Analysis section.

Suppose it is necessary to find out whether there is a relationship between the diameter of the body and the nucleus of a nerve cell. For this purpose, appropriate measurements were made in 12 cells. The data obtained are shown in Figure 74.

	1 диаметр нейрона	2 диаметр ядра нейрона
1	18	3
2	14	2
3	16	2,5
4	15	2,5
5	19	4
6	21	4
7	22	5
8	19	3
9	23	5
10	20	5

Figure 74. Example of data layout for correlation analysis

To calculate the Pearson correlation coefficient, you need to:

1. Start from the menu **Statistics / Basic Statistics/Tables /** Correlation Matrices (Figure 75).

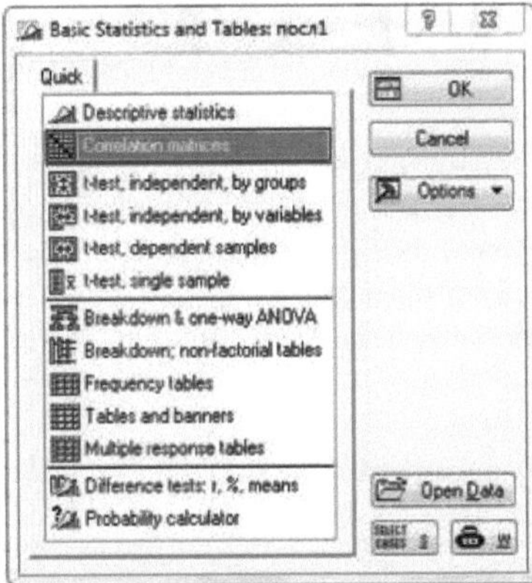

Figure 75. Correlation matrix selection dialogue box 2. Select the variables to be analysed. To do this, click the **One variable list** or **Two lists (rect. matrix)** button. In the first case, the variables are selected from one list, and in the second case - from two lists (Fig. 76).

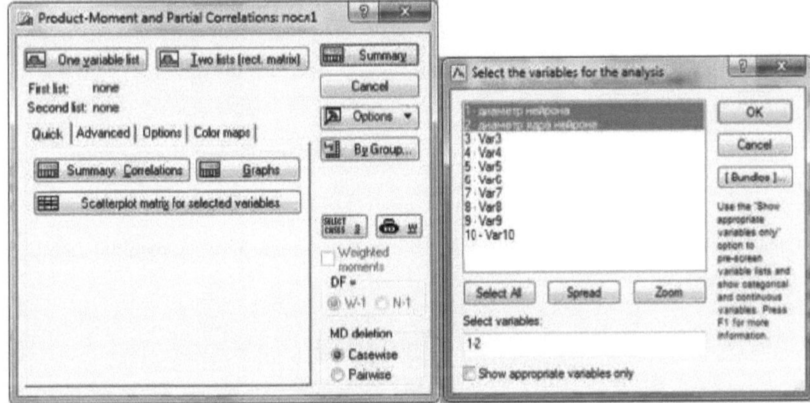

Figure 76. Dialogue windows for selecting variables, the connection between which should be checked

3. click on the button **Summary: Correlation** matrix. A table containing the calculated correlation coefficients will appear (Figure 77).

Variable	Means	Std.Dev.	диаметр нейрона	диаметр ядра нейрона
	Correlations (Spreadsheet5) Marked correlations are significant at p < ,05000 N=10 (Casewise deletion of missing data)			
диаметр нейрона	18,70000	2,983287	1,000000	0,916636
диаметр ядра нейрона	3,60000	1,149879	0,916636	1,000000

Figure 77. Result of Pearson's coefficient calculation In our case the correlation coefficient is positive and very high (r = 0.92). This indicates a direct and very high degree of correlation between the diameter of the cell body and the diameter of its nucleus. In addition to

calculating the correlation coefficient, the programme also evaluates its statistical significance
. Statistically significant correlation coefficients are highlighted in red colour (P < 0.05).

Spearman's correlation coefficient.

Suppose, when calculating the Pearson correlation coefficient for the diameter of the neuron body and its nucleus, it turned out that the values of these features do not obey the law of normal distribution. Application of the Pearson correlation coefficient in such a situation will lead to conclusions that do not correspond to reality. In this case it is necessary to use one of the nonparametric correlation coefficients. Spearman's rank correlation coefficient is one of them.

Consider its application:

1. From the **Statistics / Nonparametrics** menu, run the **Correlations** (Spearman, **Kendall tau, gamma)** module (Figure 78).

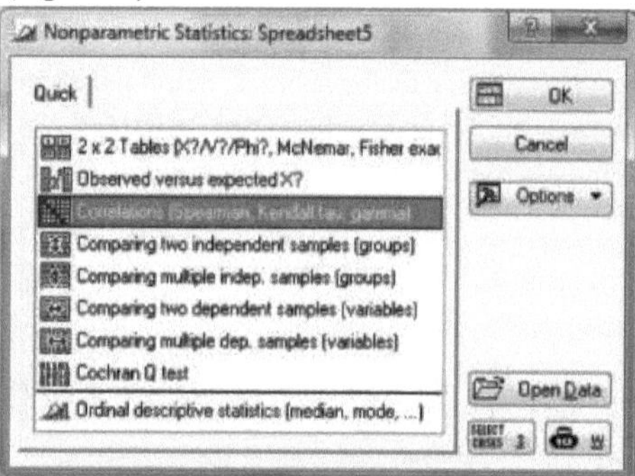

Figure 78. Correlation matrix selection dialogue window

2. Click on the **Variables** button and select the columns containing the required data (Fig. 79).

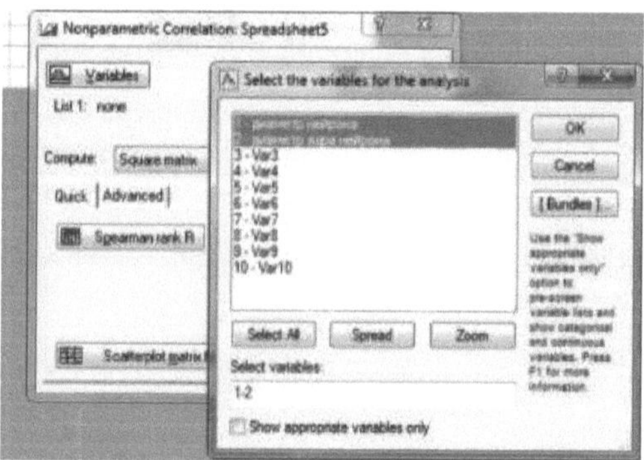

Figure 79. Dialogue window for selecting the variables, between which it is necessary to check

3. Press the **Spearman R** or **Spearman rank R** button. A table with the analysis results will appear (Fig. 80).

| Variable | Spearman Rank Order Correlations (Spreadsheet5) MD pairwise deleted Marked correlations are significant at p <.05000 | |
	диаметр нейрона	диаметр ядра нейрона
диаметр нейрона	1.000000	0.938051
диаметр ядра нейрона	0.938051	1.000000

Figure 80. Result of Spearman correlation coefficient calculation

As can be seen, the Spearman coefficient was even higher than the previously calculated Pearson coefficient.

Coefficient of association (relatedness).

Application of Pearson and Spearman correlation coefficients is possible only if the studied attributes are quantitative in nature. However, in biology, attributes are very often qualitative in nature (sex, colouring, shape, condition, etc.). Classical correlation analysis is not possible in such cases. However, for such traits it is also possible to calculate the degree of relatedness. The coefficient of association or relatedness φ (phi) allows us to do this. The closer this coefficient is to 1, the stronger the relationship.

Let us consider the application of this assay on an immunological example. A group of 111 rats was divided into groups of 57 and 54 animals. The first group was infected with pathogenic bacteria, followed by antibody injection. The animals of the second group were also infected, but antibodies were not administered (this group is the control group).

After the incubation period, the dead and surviving animals in both groups were counted. A total of 38 animals died and 73 survived. In the first group, 13 animals died and in the second group, 25 animals died. It is necessary to understand whether there is a relationship between antibody administration and animal survival? The data obtained should be presented in the form of a 2x2 *contiguity table* (a four-field table, or a table with two inputs) Table 1.

Table 1: Contiguity table

	perished	survived
Bacteria + serum	13	44
Bacteria	25	29

1. From the **Statistics / Nonparametrics** menu, start the **2X2 Tables** quadratic table analysis module (Fig. 81).

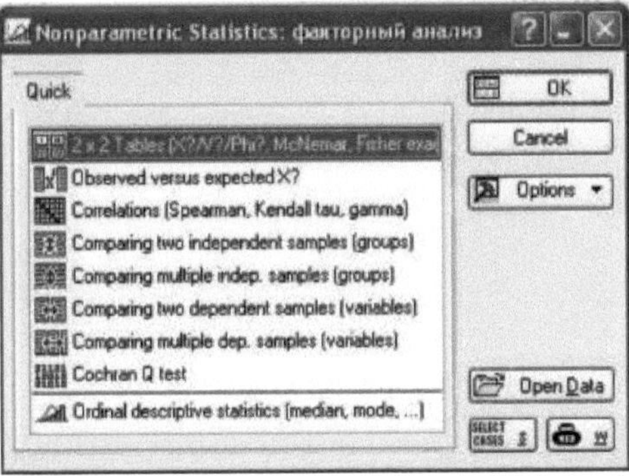

Figure 81. Dialogue window for selecting the association coefficient calculation

2.   Enter the number of animals in each of the experimental groups according to the above table (Fig. 82).

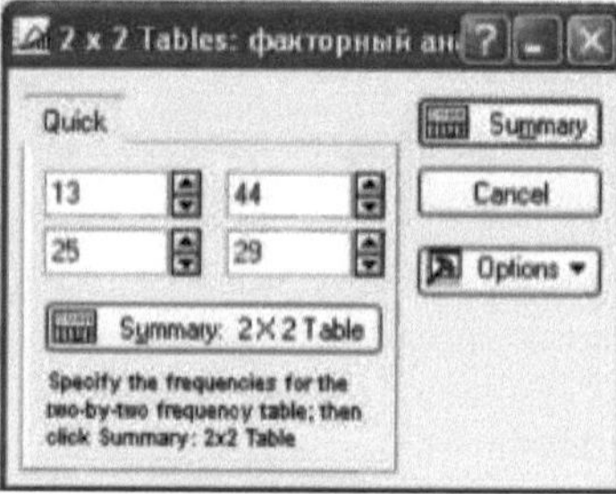

Figure 82. Module for analysing contingency tables

3.   Press the button **Summary**. As a result, a table containing a set of statistical indicators will appear (Fig. 83).

| | 2 x 2 Table (факторный анализ) | | |
	Column 1	Column 2	Row Totals
Frequencies, row 1	13	44	57
Percent of total	11,712%	39,640%	51,351%
Frequencies, row 2	25	29	54
Percent of total	22,523%	26,126%	48,649%
Column totals	38	73	111
Percent of total	34,234%	65,766%	
Chi-square (df=1)	6,80	p= ,0091	
V-square (df=1)	6,73	p= ,0095	
Yates corrected Chi-square	5,79	p= ,0161	
Phi-square	,06122		
Fisher exact p, one-tailed		p= ,0078	
two-tailed		p= ,0102	
McNemar Chi-square (A/D)	5,36	p= ,0206	
Chi-square (B/C)	4,70	p= ,0302	

Figure 83. Results of the analysis of the contingency table

We are interested in the Phi-square line. To give this coefficient a positive value, it is squared. In our case phi-square is equal to 0.061. After extracting the root we get $\varphi = 0.247$ (this operation should be done independently, it is not provided in this module). Thus, the correlation between antibody administration and survival of infected animals is rather weak.

In addition to the association coefficient, this table shows the values of the *Chi-square test ($\chi 2$ test)*. This test tests the null hypothesis that the relationship between antibody administration and animal survival is random. Since the probability of error, rejecting this assumption, is less than 0.05 ($P = 0.0091$), we can conclude that despite the weak effect of antibody administration, the survival of animals in the control and experimental groups is still statistically significantly different.

CHAPTER 5

Regression analysis.

Regression analysis, along with correlation analysis, is one of the most common methods of processing experimental data in the study of dependencies. The essence of this analysis is to determine to what extent the change in one variable (dependent variable) is caused by the influence of one or more independent variables (factors).

Since regression analysis belongs to the group of parametric methods of statistical analysis, its application requires fulfilment of a number of obligatory conditions:

1. linear nature of the dependence;

2. "Normal" distribution of data.

Let us perform regression analysis on the example of systolic blood pressure (SBP) indices in people of different ages.

Firstly, let us enter the data obtained during the study into the table (Figure 84)

	1 возраст	2 давление мм. рт. ст.	3 Var3
1	30	108	
2	30	110	
3	40	125	
4	40	120	
5	40	118	
6	50	132	
7	50	137	
8	50	134	
9	60	148	
10	60	151	
11	60	146	
12	60	147	
13	70	162	
14	70	156	
15	70	164	
16	70	159	

Figure 84. Example of data design for regression analysis

Regression analysis can be performed in several modules of the STATISTICA programme. Let's use the Multiple Regression **Analysis** module.

1.   Start the corresponding module from the menu: **Statistics / Multiple** Regression Analysis (Figure 85).

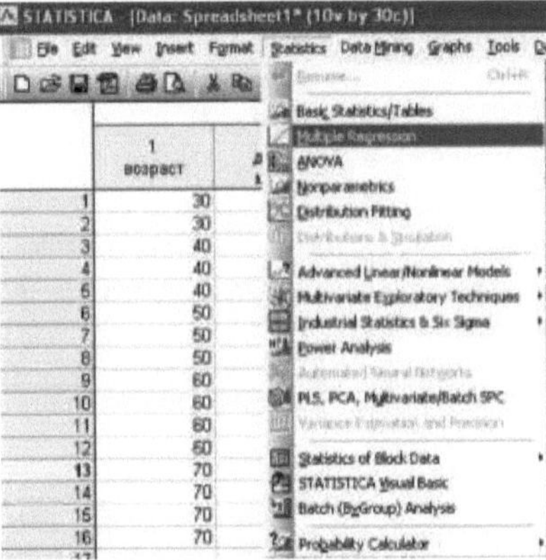

Figure 85. Dialogue window for selecting regression analysis

2.   Click on the **Variables** button and specify the Dependent variable and Independent variable (in our case, CAD depends on age) (Fig. 86).

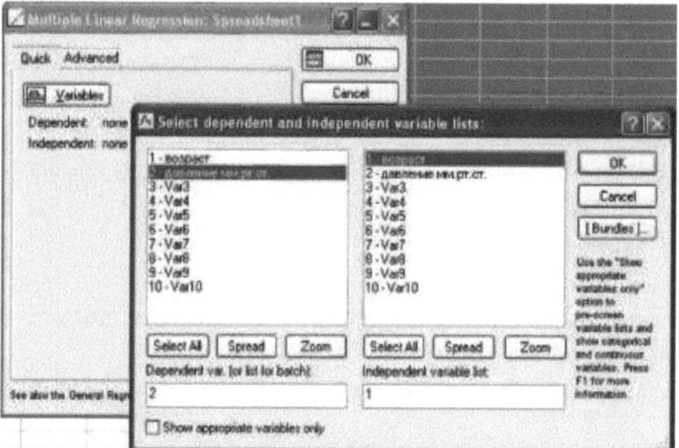

Figure 86. Dialogue window for variable selection

3.   Press the **OK** button. As a result, a window with preliminary results of the analysis will appear (Fig. 87).

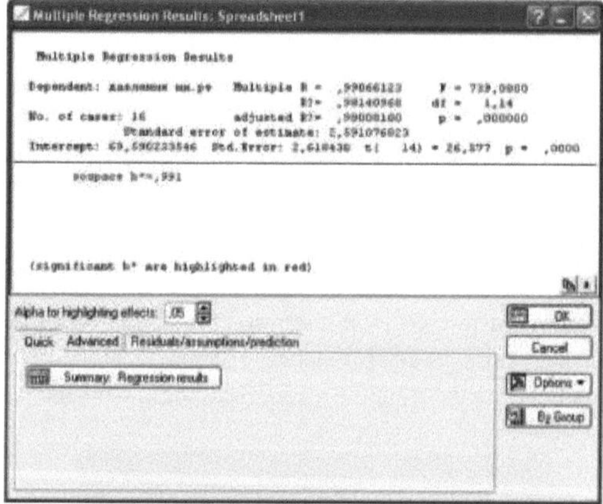

Figure 87. Preliminary results of regression analysis

The most important indicator in the table is the coefficient *R2* or coefficient of determination. It reflects the "quality" of the calculated regression. In our case $R2 = 0.98$, therefore, changes in the dependent variable (CAD) are explained by 98% by changes in the independent factor or variable (age). We can say that the constructed regression model perfectly describes the relationship between age and blood pressure;

4.   Click on the **Summary:** Regression **results** button. A table with the following results will appear (Figure 88):

N=16	Regression Summary for Dependent Variable: давление мм.рт.ст. (Spreads R= ,99066123 R?= ,98140968 Adjusted R?= ,98008180 F(1,14)=739,08 p<,00000 Std.Error of estimate: 2,5911					
	b*	Std.Err. of b*	b	Std.Err. of b	t(14)	p-value
Intercept			69,59023	2,618438	26,57700	0,000000
возраст	0,990661	0,036440	1,29830	0,047756	27,18603	0,000000

The table shows that both regression coefficients are significantly different from 0 (P << 0.001).

However, in biology, it is not the regression itself that is of most interest, but the effect that one variable has on another. In our case, it is important to know how CAD will change with age.

For this purpose, it is necessary:

1. In the **Multiple** Regression **Result** window, select the Residuals / assumptions / **prediction** tab and click the **Predict dependent variable** button (Figure 89).

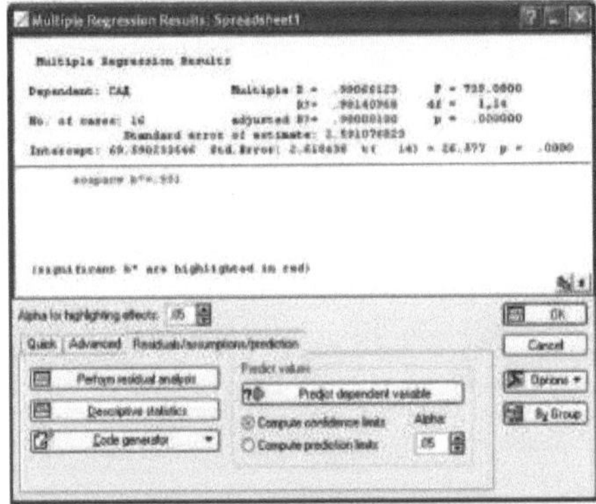

Figure 89. Preliminary results of regression analysis

2. In the window that appears, enter the data for the forecast. Let us assume that we need to know what values of CAD can be expected in persons aged 80 years (Fig. 90).

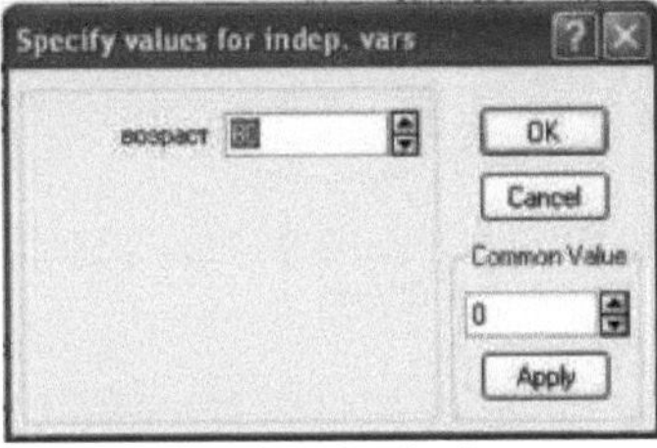

Figure 90. Dialogue window for entering data for the forecast

3. Press the **OK** button. As a result, a window will appear, in which the prediction results are presented (Fig. 91.).

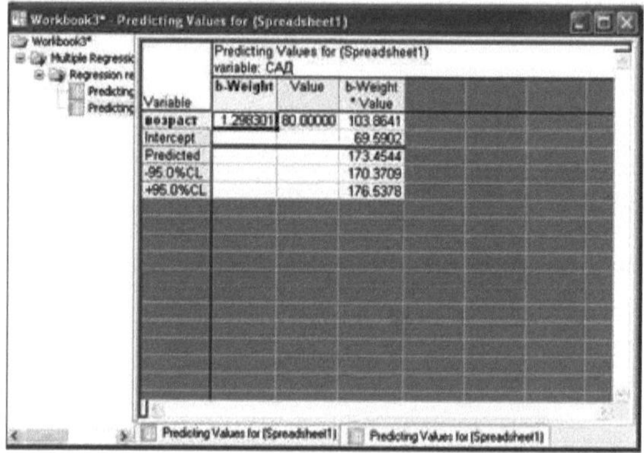

Figure 91. Results of regression analysis

In the table, we are interested in the **Predicted** row. As can be seen, by the age of 80 the value of CAD will reach 173 mm. Rt. St.

A significant part of regression analysis is the *analysis of residuals* (the difference between the observed values of the dependent variable and those values predicted by the regression model).

To perform this analysis, it is necessary to:

1.   Select the Residuals / **Assumptions** / Predictions tab. Click the **Perform residual analysis** button (Figure 92).

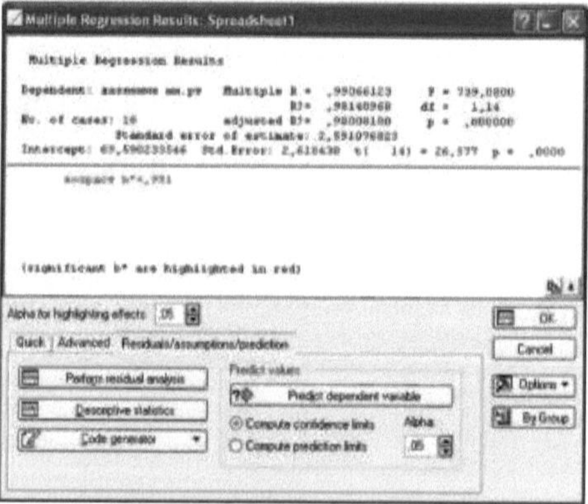

Figure 92. Residuals analysis dialogue window

2.   First: check the "normality" of the distribution of residuals. On the **Quick** tab, click the **Normal plot of residuals** button (Figure 93).

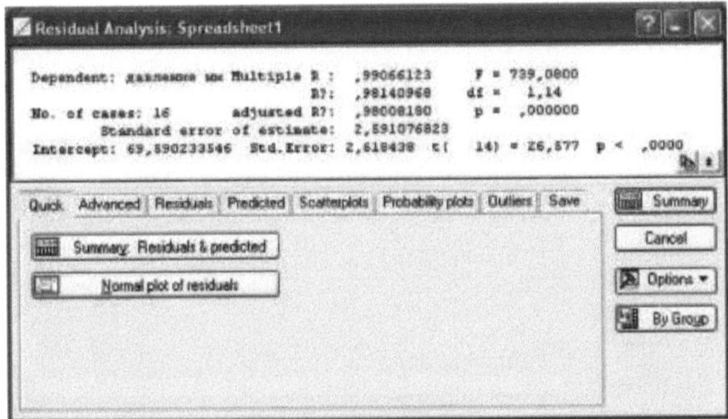

Figure 93. Dialogue window for evaluation of residuals for conformity to the law of normal distribution

3. As a result, a graph of normal probabilities will be plotted (Fig. 94). If the points on this graph are compactly located along the theoretically expected line, the residuals are distributed "normally", the application of the linear regression model is correct. If this condition is not fulfilled, application of linear regression is impossible. The solution in this case can be data transformation (transformation methods are described below).

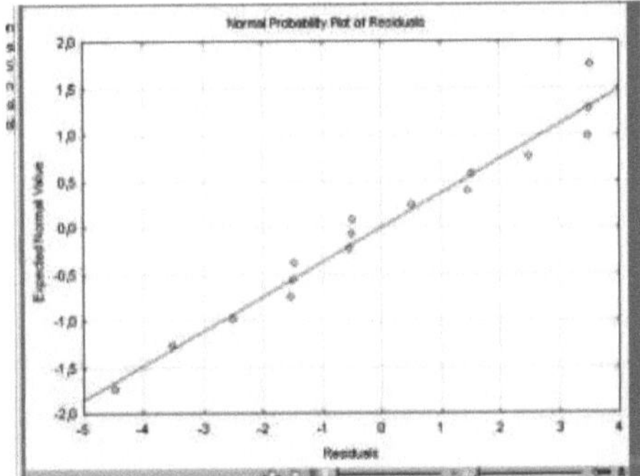

Figure 94. Graph of normal probabilities of residuals

4. Second: check the variance of the residuals. The variance should remain unchanged over the entire range of values of the analysed variables. On the Scatterplots tab, click the **Predicted vs. Residuals button** (Figure 95).

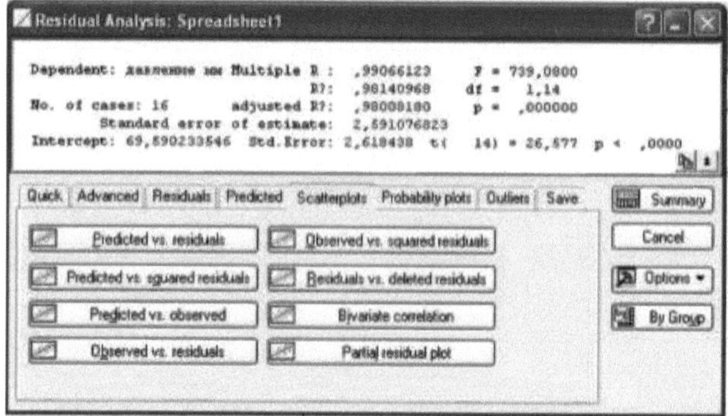

Figure 95. Dialogue window for estimating the variance of residuals

As a result, a graph of the dependence of residual values on the values of the dependent variable predicted by the model will be plotted (Fig. 96).

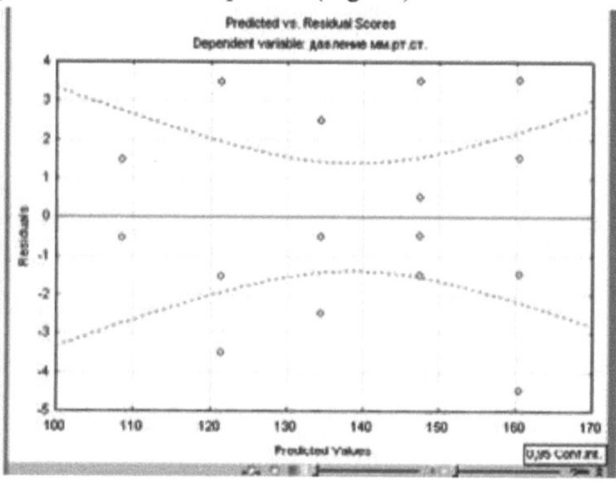

Figure 96. Graph of dependence of residuals values on the values of the dependent variable predicted by the model

If the variance is unchanged (the condition is fulfilled), the points on this graph will be arranged chaotically. If there is any regularity in the arrangement of points (the points are grouped to the left or right, stacked along a straight line, etc.), linear regression analysis cannot be used. Data transformation can also be a solution.

In our case, both conditions are fulfilled, which confirms the correctness of the calculated regression model.

Transformation of non-linearly related features

A very serious limitation for the application of regression analysis in biology is the non-linear nature of relationships between many biological traits. For example, the relationship between body size and the intensity of metabolic processes is nonlinear (steppe or exponential). In such a situation, a certain transformation of the initial data can help. This allows to translate them into a different scale of measurement and thus "equalise" the non-linearity.

Fig. 97 presents data on the intensity of respiration processes and body size of *Daphnia magna*.

	1 длинна тела мм.	2 интенсивн. дыхания
1	0,556	0,018
2	0,6	0,021
3	0,694	0,035
4	0,779	0,13
5	0,849	0,055
6	0,954	0,28
7	1,099	0,48
8	1,102	0,54
9	1,204	0,72
10	1,205	0,731
11		

Figure 97. Example of data design for regression analysis

The nature of the relationship between two variables should be checked even before conducting regression analysis. For this purpose it is necessary to:

1. On the **Graphs** tab, select the Scatterplots section.

As a result, a scattering diagram will be plotted (Figs. 98, 99).

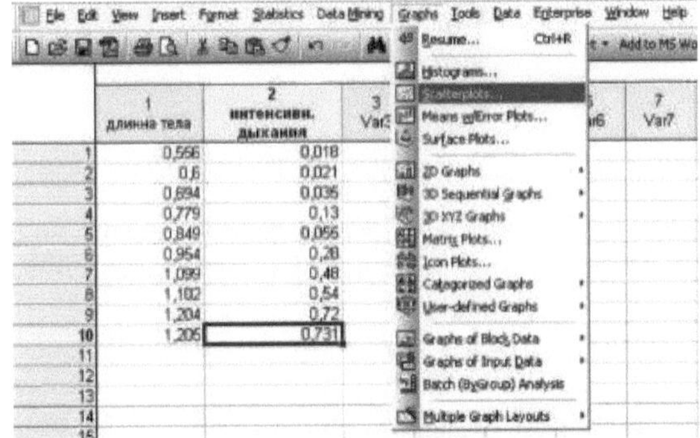

Figure 98. Dialogue window for drawing the scattering diagram

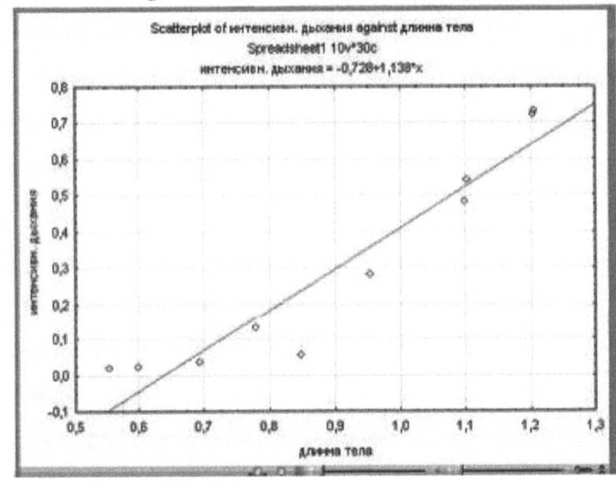

Figure 99. Scatter diagram for body size and respiration rate data

The arrangement of points on the diagram shows that the relationship between daphnia body size and metabolic rate is not linear. Linear regression analysis is not applicable. However, this problem can be solved by logarithmising the values of one or (more often) both of the analysed traits.

This transformation can be done by assigning so-called *Long names* to variables in the form of formulas.

2. Let's prologue column 1, containing the values of daphnia body size. To do this, click twice on the header of any free column (for example, column Var 3). This will cause the window of variable properties settings to appear (Fig. 100).

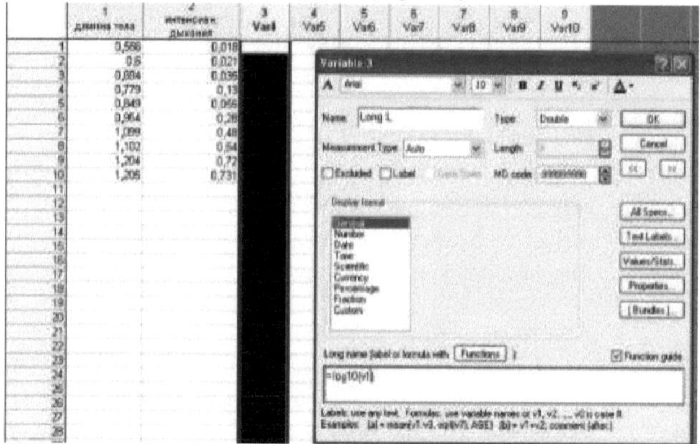

Figure 100. Dialogue window for assigning long names to variables

3. In the *Long name* field enter the formula = *log* 10(*v* 1), where *v* 1 is the column with data on the length of the crayfish. In the *Name field,* enter a short name for the variable, for example, "*Log L*".

4. Click the **OK** button. The panel **"Expression OK. Recalculate** the variable **now?"** **panel appears**. Click **Yes**. The programme will transform the values of the first column and they will appear in column 3 (Fig. 101).

	1 длинна тела	2 интенсивн. дыхания	3 Long L
1	0,556	0,018	-0,25493
2	0,6	0,021	-0,22185
3	0,694	0,035	-0,15864
4	0,779	0,13	-0,10846
5	0,849	0,055	-0,07109
6	0,954	0,28	-0,02045
7	1,099	0,48	0,040998
8	1,102	0,54	0,042182
9	1,204	0,72	0,080626
10	1,205	0,731	0,080987
11			

Figure 101. Results of logarithmic transformation of experimental data

5. A similar operation should be performed for the data on the intensity of metabolic processes. Note, it is necessary to enter the formula =*log10*(*v2*) as its long name (Fig. 102).

	1 длинна тела	2 интенсивн. дыхания	3 Long L	Intensity of breathin
1	0,556	0,018	-0,25493	-1,74473
2	0,6	0,021	-0,22185	-1,67778
3	0,694	0,035	-0,15864	-1,45593
4	0,779	0,13	-0,10846	-0,88606
5	0,849	0,055	-0,07109	-1,25964
6	0,954	0,28	-0,02045	-0,55284
7	1,099	0,48	0,040998	-0,31876
8	1,102	0,54	0,042182	-0,26761
9	1,204	0,72	0,080626	-0,14267
10	1,205	0,731	0,080987	-0,13608

Figure 102. Results of logarithmic transformation of experimental data

6. If we plot a scatter plot for the transformed data, we can see that the points fit along a straight line much more compactly (Fig. 103), and it is possible to apply ordinary linear regression.

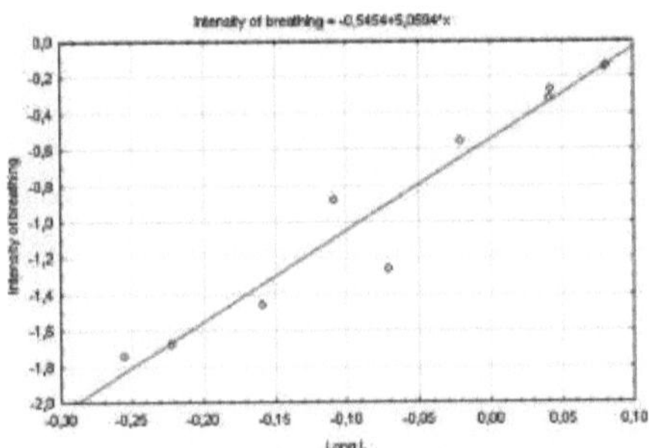

Figure 103. Scatter diagram for body size and respiration rate data after the logarithmisation procedure

It is important to note that the procedure of data transformation is applicable not only for "levelling" non-linear relationships between traits. Logarithmisation allows to "approximate" the distribution of data to "normal", as well as to achieve homogeneity of dispersion in groups. All this allows the use of more powerful parametric methods of data analysis.

CHAPTER 6

Cluster analysis.

The main task solved by cluster analysis is to divide the initial set of studied objects and attributes into homogeneous, in some sense, groups or clusters.

Cluster analysis methods allow to solve the following problems:

1. The classification of objects with reference to various characteristics or attributes reflecting their nature;

2. Testing assumptions or searching for some internal structure in a set of objects under study.

The advantage of cluster analysis is that it belongs to the group of non-parametric methods of statistical analysis. Its application is possible in small groups or when the requirements of "normality" of data distribution are not fulfilled.

The following varieties of cluster analysis are implemented in the STATISTICA package:

1. Hierarchical algorithms or tree clustering;
2. K-means method;
3. Two-entry merger.

Hierarchical algorithms or tree clustering.

The purpose of this type of cluster analysis is to group objects into sufficiently large groups (clusters) based on similarity or "distance" between objects. The result of such clustering is the construction of a hierarchical tree. The algorithm of this type of cluster analysis consists in successive grouping of the closest and then more and more distant objects into groups.

Let us consider this type of analysis on the example of an ecological study of different plant communities. Nine different phytocenotic parameters were measured within plant communities of different species composition (Fig. 104). It is necessary to determine which communities are most similar in structural characteristics and which have the greatest differences. The data obtained in the course of the study are included in the table (Fig. 104).

	2 диаметр\древостой	3 сомкнутость крон\древостой	4 высота\подрост	5 диаметр\подрост	6 высота\подлесок	7 диаметр\подлесок	8 высота\травостой	9 покрытие\травостой
дубрава остепненная	63.9	11.9	1.9	2.6	0.7	0.6	0.32	25.8
дубрава мятликовая	60.6	29.8	1.3	1.1	1.1	0.8	0.37	30
дубрава разнотравная	64.3	23.6	2.95	2.2	1.9	1.6	0.47	45.9
дубрава ландышевая	62.3	25.5	2.1	1.9	1.5	1.2	0.62	75.3
липо-дубрава ландышевая	56.1	58.4	2.1	2.6	0.9	0.7	0.6	63.4
липо-дубрава крапивная	62.4	36.4	2.2	2.4	0.9	1	0.3	32.7
липняк снытевый	54.9	62.4	3.1	2.59	1.2	1.5	0.58	66.4
липо-кленовник	48.9	70	3.5	2.9	1.9	1.1	0.25	32.5
березняк	50.1	62.5	3	1.9	1.6	1.2	0.24	70.8
осинник	49.3	30.2	1.5	1.2	1.1	0.3	0.36	60.2
сосняк	49.3	30	0.5	0.3	1.1	0.3	0.25	5.01

Figure 104. Example of data design for performing tree clustering

1. Start the cluster analysis module **Statistics / Multivariate Exploratory / Cluster Analysis** (Statistics / Multivariate Research Methods / Cluster Analysis) (Fig. 105).

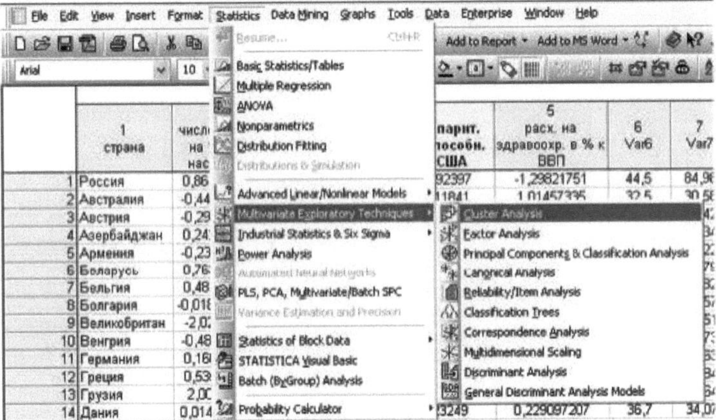

Figure 105. Dialogue window for selecting the cluster analysis method

2. In the list of methods, select **Joining (tree** clustering) and click **OK** (Fig. 106).

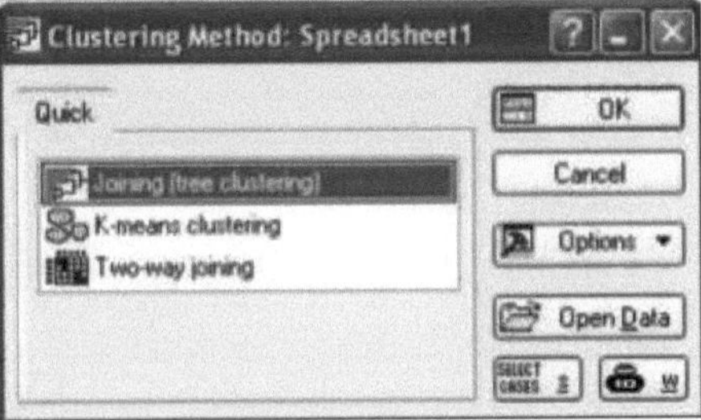

Figure 106. Dialogue window for selecting the cluster analysis method

3. Click the **Variables** button and select the variables to be analysed (for clustering). Click **OK** (Fig. 107).

4. Since the indicators characterising plant communities are arranged in **rows** and have not been transformed, select the menu item **Cases (rows) in the Cluster row,** and select **Raw data** in the **Input file row** (Figure 107).

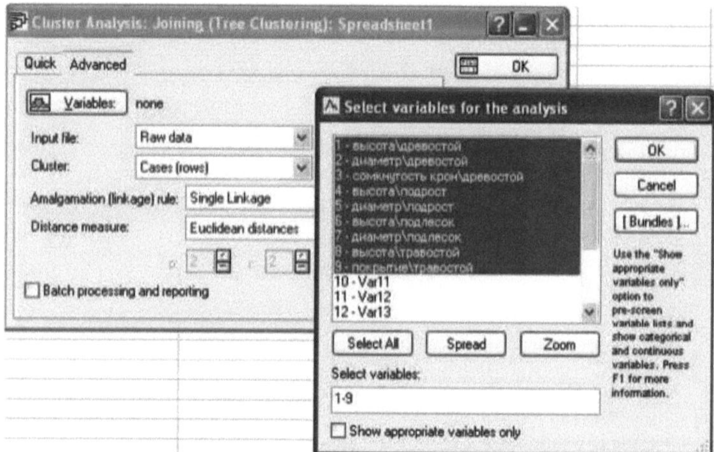

Figure 107. Dialogue window for selecting cluster analysis settings

5.   After clicking **OK, a** dialogue box appears, in which you need to select the tree diagram type (Fig. 108). Select a convenient type of diagram to present the results, we get the result (Fig. 109).

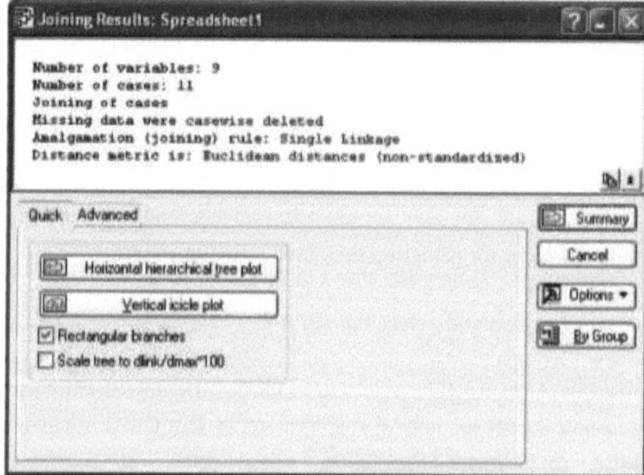

Figure 108. Dialogue window for selecting cluster analysis settings

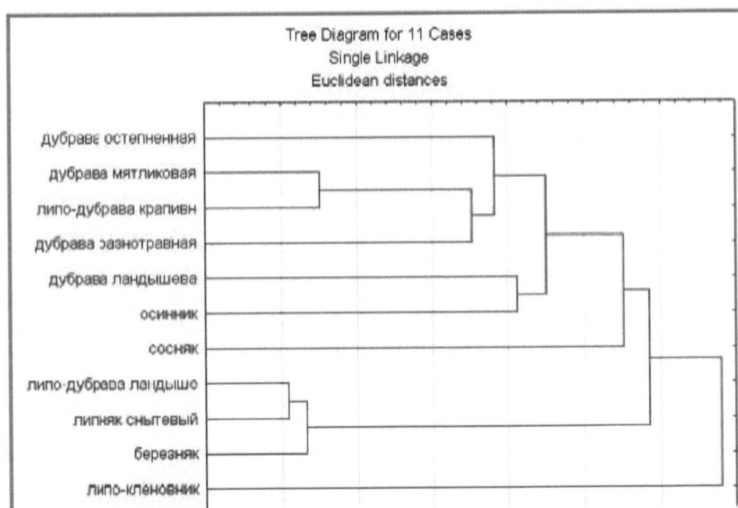

Figure 109. Display of cluster analysis results

The resulting cluster diagram shows the association of plant communities into groups based on the similarity of phytocenotic parameters. Two clusters are clearly visible on the diagram. The first one is formed by steppe oak, bluegrass, and lipo-krapivy. The second cluster includes lily-of-the-valley lipo-oak, snythe limber, lipo-maple and birch forest. Aspen forest and lily-of-the-valley oak are similar to the communities of the first and second groups. The pine forest is the most different from all other communities.

K-means method.

It is one of the most frequently used methods of cluster analysis. This method allows to divide a set of objects into a given number of clusters **K.** These clusters are located at maximum distances from each other. Once the clustering results are obtained, the averages for each cluster are calculated for each dimension, and it is assessed how different the clusters are from each other. Ideally, for most dimensions, you should get significantly different averages. The F-statistic values obtained for each dimension are another indicator of the quality of cluster discrimination.

It is very important to remember that when applying this method of analysis, the researcher must have a hypothesis (assumption) about the number of possible clusters. This method builds exactly as many clusters as the researcher specifies. The formed clusters will be located at the largest possible distances from each other.

For example, let us consider the results of the study of countries by some indicators of living standards of the population. Based on these data, it is necessary to divide the countries into groups. Differences between the groups should be maximised and within the groups minimal (countries should be as similar as possible). The results of the study and an example of their design are shown in Figure 110.

	1 число врачей на 10 тыс. населения	2 смертность на 100 тыс. населения	3 ВВП по парит. покуп. способн. в % к США	4 расх. на здравоохр. в % к ВВП
Россия	44.5	84.98	20.4	3.2
Австралия	32.5	30.58	71.4	8.5
Австрия	33.9	38.42	78.7	9.2
Азербайджан	38.8	60.34	12.1	3.3
Армения	34.4	60.22	10.9	3.2
Беларусь	43.6	60.79	20.4	5.4
Бельгия	41	29.82	79.7	8.3
Болгария	36.4	70.57	17.3	5.4
Великобритания	17.9	34.51	69.7	7.1
Венгрия	32.1	64.73	24.5	6
Германия	38.1	36.63	76.2	8.6
Греция	41.5	32.84	44.4	5.7
Грузия	55	62.64	11.3	3.5
Дания	36.7	34.07	79.2	6.7
Ирландия	15.8	39.27	57	6.7
Испания	40.9	28.46	54.8	7.3
Италия	49.4	30.27	72.1	8.5
Казахстан	38.1	69.04	13.4	3.3
Канада	27.6	25.42	79.9	10.2
Киргизия	33.2	63.13	11.2	3.4

Figure 110. Example of data design for cluster analysis

Because the variables used for analysis have different units of measurement (or if the scales of measurement do not match sharply), prior **normalisation is** necessary. Normalisation is the conversion (transformation) of raw data into dimensionless values. For this purpose it is necessary:

1. Right-click on the variable name . In the opened window select the sequence of commands: Fill / Standardize Block / Standardize <u>Columns</u> (Fig. 111).

Figure 111. Dialogue window for normalisation of variables

The values of the normalised variable will become zero and the variance will become one. The same operation should be done with all variables. After that it is possible to proceed with the cluster analysis.

2. Start the **Statistics / Multivariate Exploratory / Cluster Analysis** module (Figure 112).

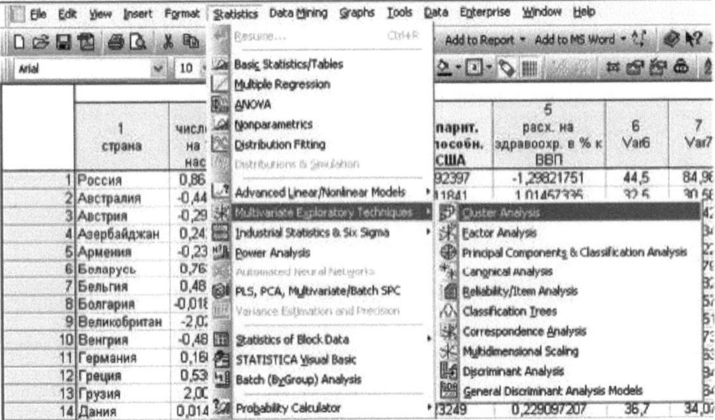

Figure 112. Dialogue window for selecting the cluster analysis method

Figure 112. Dialogue window for selecting the method of cluster analysis 2. In the list of methods, select **k-means clustering** and click **OK** (Fig. 113).

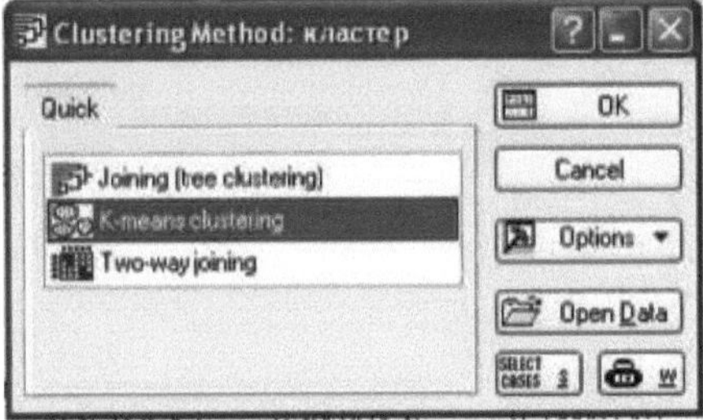

Figure 113. Dialogue window for selecting the cluster analysis method

3.   Click the **Variables** button and select the variables to be analysed (for clustering). Click **OK** (Fig. 114).

4.   Since the living standards indicators are arranged in **rows**, in the **Cluster** field, select **Cases (rows)**. In the **Number of clusters** field, set the number of clusters into which the sample should be divided, for example, 3. In the **Number of (iterations)** line, set the maximum number of iterations used in the construction of classes, for example, 10 (Figure 114).

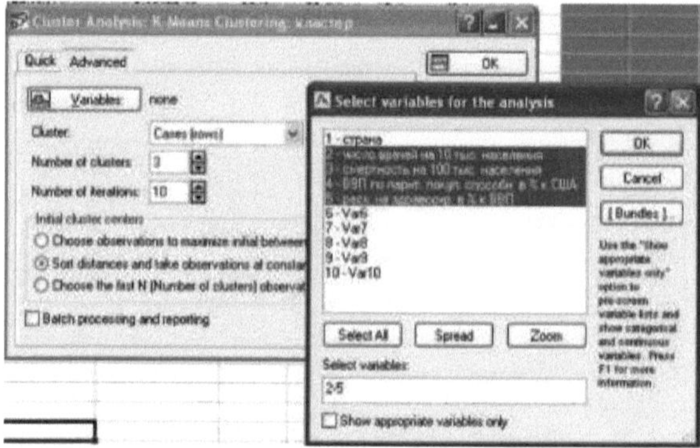

Figure 114. Dialogue window for selecting cluster analysis settings

After clicking **OK**, a dialogue window with clustering results will appear (Fig. 115.. The results window at the top provides the following information:

Number of variables (Number of variables) - 4;

Number of registers (Number of cases) - 20;

K-means clustering of cases - k-means clustering method;

Number of clusters - 3;

Solution was obtained after 2 iterations - The solution was found after 2 iterations.

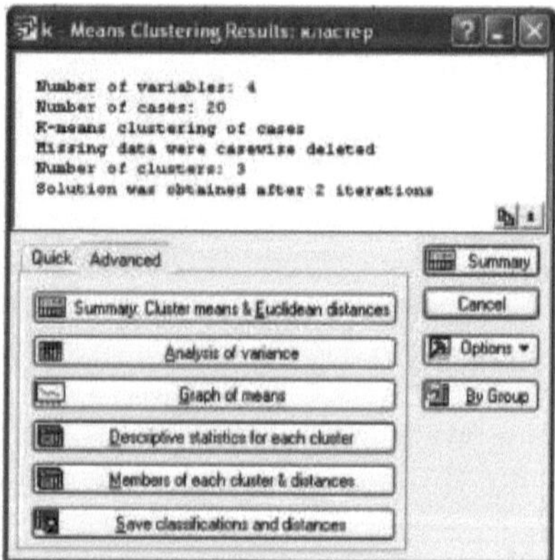

Figure 115. Dialogue window with clustering results

6. Select the **Advanced** tab. Use the buttons in this window to view the results of the analysis.

The function button Cluster Means&Euclidean Distances displays 2 tables. The first one shows the average values for each cluster (averaging is done within a cluster). The second one shows Euclidean distances and squares of Euclidean distances between clusters (Fig. 116).

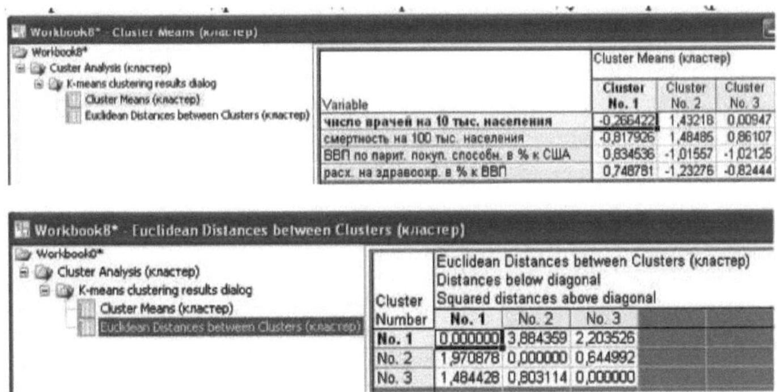

Figure 116. Tables of results Cluster averages & Euclidean distances

Functional button Analysis of variance allows to view the table of variance analysis, where, for example, the sums of squares of deviation of objects from the centres of clusters (SS Within) and sums of squares of deviations between the centres of clusters (SS Between), values of F-statistics, significance levels p are displayed (Fig. 117).

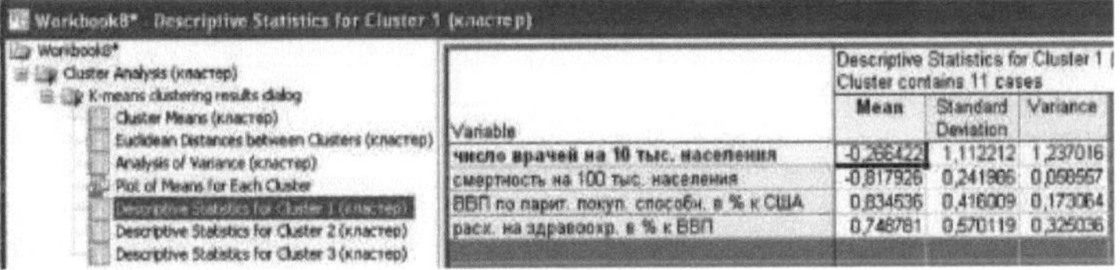

Figure 117. Table of results of analysis of variance

Functional button Descriptive Statistics for each clusters. It displays tables with descriptive statistics for each cluster (mean, standard deviation, variance) (Figure 118).

Figure 118. Table with descriptive statistics for each cluster

Function button Graph of means. Displays the average values for each cluster in a line graph. The curves in this graph correspond to the selected clusters. On the horizontal axis are plotted the variables included in the

analyses. On the vertical axis are the average values for the countries included in each (Figure 119).of the clusters

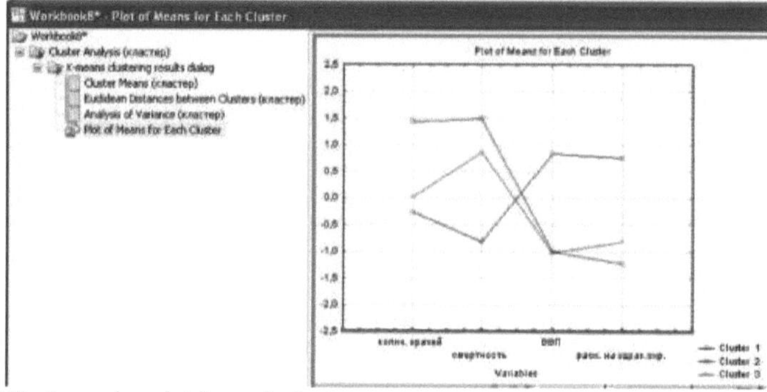

Figure 119. Averaging plot for each cluster

Function button Member of each cluster & distances. Shows the distribution of countries in their respective clusters. The displayed tables show the numbers of countries assigned to one or another cluster. In the header of the table you can see that the first cluster contains 11 countries, their numbers are listed in the **Case #** column (2, 3, 7, 9, 11, 12, etc.). The rows show the distances from each country to the centre of the cluster (Figure 120).

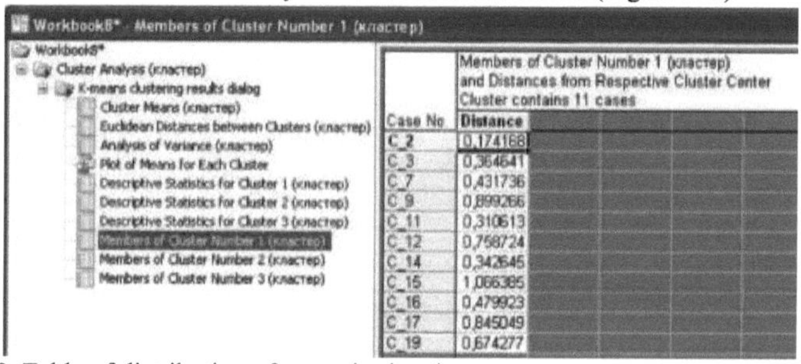

Figure 120. Table of distribution of countries by clusters

The Save classifications and distances **function button** displays the numbers of objects included in each cluster and the distances of objects to the centre of each cluster. Information about the belonging of objects to clusters can be written to a file and used in further analyses (Figure 121).

	кластер						
	1 число врачей на 10 тыс. населения	2 смертность на 100 тыс. населения	3 ВВП по парит. покуп. способн. в % к США	4 расх. на здравоохр. в % к ВВП	5 CASE_NO	6 CLUSTER	7 DISTANCE
C_1	0,861698627	2,1113567	-0,858292397	-1,29821751	1	2	0,43
C_2	-0,442259045	-0,93984172	0,904611841	1,01457335	2	1	0,17
C_3	-0,290130626	-0,500110327	1,15694911	1,32003629	3	1	0,36
C_4	0,242318838	0,729342751	-1,14519642	-1,25457994	4	3	0,26
C_5	-0,235799048	0,722612168	-1,18667652	-1,29821751	5	3	0,29
C_6	0,763901986	0,754582435	-0,858292397	-0,338191115	6	3	0,46
C_7	0,481377781	-0,982468743	1,19151586	0,927298219	7	1	0,43
C_8	-0,0184727365	1,30312491	-0,965449321	-0,338191115	8	3	0,33
C_9	-2,02874112	-0,719415142	0,845848367	0,40364746	9	1	0,90
C_10	-0,485724307	0,975569696	-0,716568723	-0,0763657357	10	3	0,48
C_11	0,166254629	-0,600508183	1,07053224	1,05821091	11	1	0,31
C_12	0,535709359	-0,813082415	-0,0288904023	-0,207278425	12	1	0,76
C_13	2,00266196	0,858345583	-1,17284982	-1,16730482	13	2	0,43
C_14	0,0141262103	-0,744093944	1,17423249	0,229097207	14	1	0,34
C_15	-2,25693375	-0,452435367	0,406850645	0,229097207	15	1	1,07
C_16	0,470511465	-1,05874868	0,330803795	0,490922587	16	1	0,48
C_17	1,39414829	-0,957229058	0,928808566	1,01457335	17	1	0,85
C_18	0,166254629	1,21730999	-1,10025965	-1,25457994	18	3	0,29
C_19	-0,974708509	-1,22925677	1,19842921	1,75641192	19	1	0,67
C_20	-0,366194835	0,324946916	-1,17630649	-1,21094238	20	3	0,39

Figure 121. Table of distribution of countries by clusters

CHAPTER 7

Calculation of sample size (volume) or power analysis.

Calculation of the sample size is one of the essential stages of planning an experiment. Deciding on the size of groups (power of the study) is necessary in order to avoid the second kind of error when analysing the obtained data. Let me remind you that the 1st kind of error is the probability of falsely rejecting the null hypothesis, i.e. finding differences where there are none. The maximum permissible probability of this error is 5% and is called the significance level. The 2nd kind of error is the probability of falsely accepting the null hypothesis, i.e. not finding differences where they exist.

In order to get an idea of the sample size, several indicators need to be known:
1. The size of the expected effect;
2. Mean values of traits or variables;
3. Standard deviation of the mean values of the investigated attributes or
of the variables (variance value).

The question arises, where to get these indicators if the study is still being planned and they are simply not known? In this case, the information needed to estimate the sample size is obtained either from the results of one's own previous research or from similar studies described in the literature. In addition, some assumptions have to be made.

Let's take one example: suppose we are conducting a clinical trial of the effects of two medicines (A and B) that affect systolic blood pressure. We have sufficient resources to enrol 25 patients in the study to test each of these drugs. Will this be enough to detect meaningful results? In other words, will our study have sufficient power?

The first question we need to answer is how large is the effect size to be detected? In other words, how much should systolic blood pressure change in patients using a particular drug? Of course, we don't know this, which is why we are doing the study! But we can make some assumptions. For example, we have results from previous studies that included drug A, and we believe that the mean blood pressure for drug B will differ by about 10% from the mean for drug A. If the mean systolic blood pressure for drug A is 120 mmHg, then the effect size would be 12 mmHg.

The second question is what is the variability of systolic blood pressure measurement? A previous study of drug A demonstrated that the standard deviation of systolic blood pressure is 10 mmHg. Assume that the standard deviation will be about the same in groups receiving either drug.

Based on these provisions, the power of the study can be calculated. 1. From the menu, launch the appropriate module: **Statistics / Power Analysis** (Fig. 122).

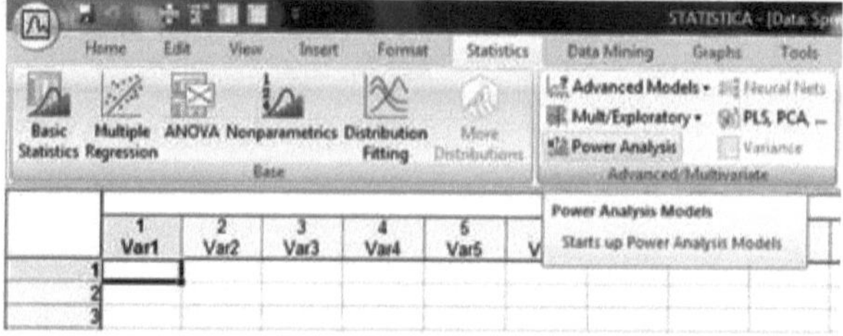

Figure 122. Window of the power analysis module

2. In the window that appears, select the criterion **Two Means, t-Test, Independent Samples** (Student's t-test for independent samples) (Fig. 123).

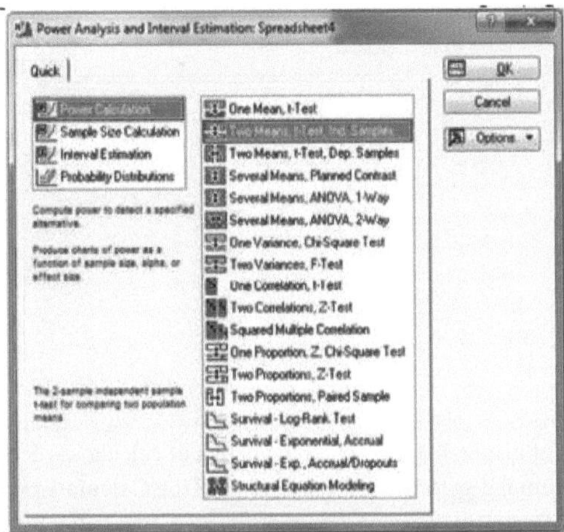

Figure 123. Dialogue window for selecting the statistical criterion 3. Click **OK** and in the next dialogue box set the known parameters. *Mu1* and
Mu2
are the known and expected mean values of blood pressure indices in the studied groups; *N1* and *N2* are the number of patients planned to be involved in the study; *Sigma* is the standard deviation in the studied groups: *Alpha is the* level of error of the first kind (ε) = 0.05 (Fig. 124).

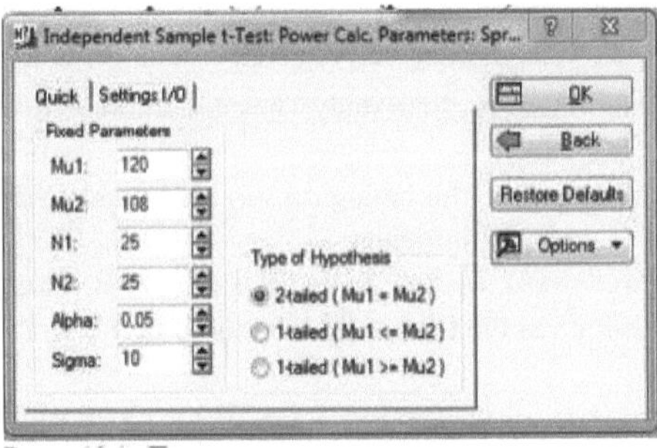

Figure 124. Dialogue window for entering known parameters
4. Press **OK**. A dialogue window appears, which displays the parameters on the basis of which the analysis is performed (Fig. 125).

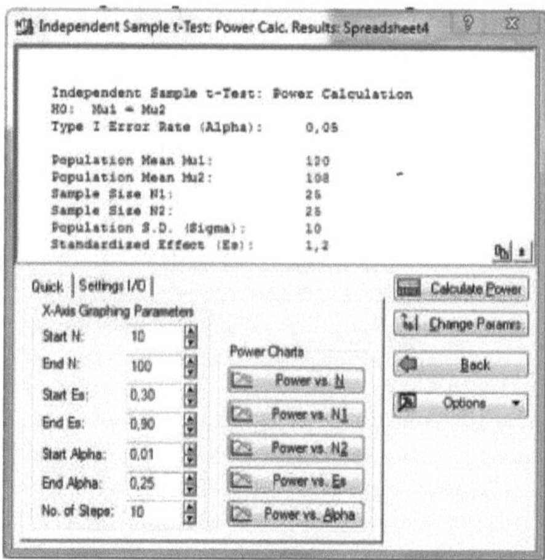

Fig. 125. The window of displaying the main parameters of calculation 5. To calculate the power taking into account the specified parameters, click the **Calculate** power button. The final table will contain the results of power estimation, as shown in the figure (Fig. 126).

	Power Calculation (Spreadsheet4) Two Means, t-Test, Ind. Samples H0: Mu1 = Mu2
	Value
Population Mean Mu1	120.0000
Population Mean Mu2	108.0000
Population S.D. (Sigma)	10.0000
Standardized Effect (Es)	1.2000
Sample Size N1	25.0000
Sample Size N2	25.0000
Type I Error Rate (Alpha)	0.0500
Critical Value of t	2.0106
Power	0.9860

Figure 126. Power estimation results of the study The table shows that for this combination of parameters, the power is 0.98.

The minimum acceptable power level for biological surveys should not be less than 0.8, i.e. the planned sample size is more than sufficient.

If the power is small (Power < 0.8), we need to understand at what value of N we will get the normal power. 6. To do this, press the **Power vs. N** button (Fig. 127).

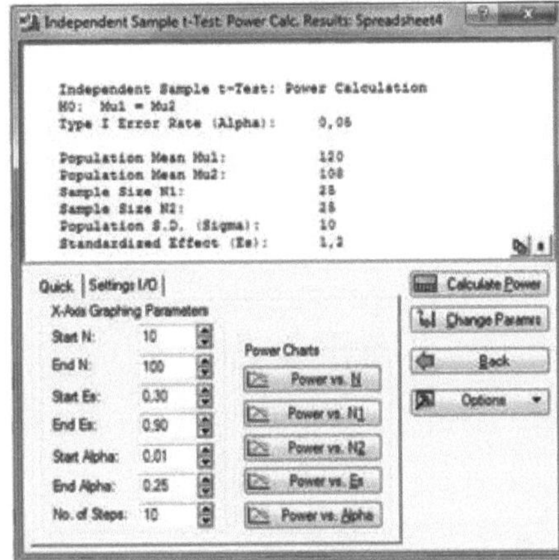

Figure 127. The dialogue window for calculating the required power of the study 7. The next window will display a graph of the ratio of sample size to power (Fig. 128). The graph shows that a power of 0.8 can be achieved with a sample size of approximately 13 people.

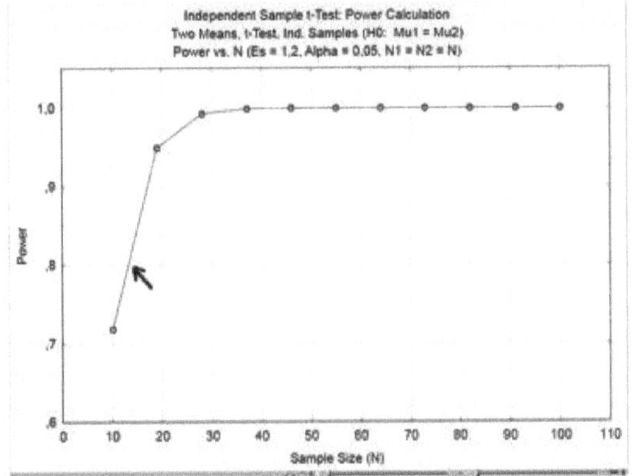

Figure 128. Graph of sample size to power ratio **Similarly, power analysis is performed for dependent samples.**

Let us turn to the data obtained in the course of observations of changes in the amount of immunoglobulins in mice "before" and "after" exercise (p. 20). Let me remind you that 19 animals were involved in the study, the mean values of the amount of immunoglobulins "before" and "after" exercise were 6.8 mm. and 7.3 mm. (diameter of the precipitation ring), standard deviation 1.2 mm. and 1.4 mm.

1. From menu launch module **Poweranalysis/ Two Means, t-Test, Dependent Samples** (Figure 129).

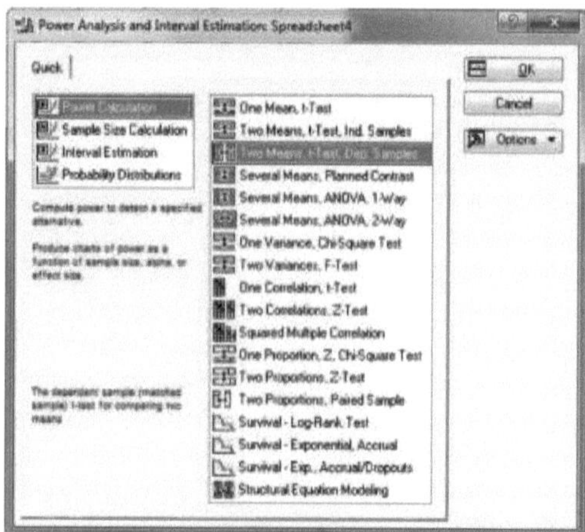

Figure 129. Dialogue window for selecting the statistical criterion

2. Click **OK** and in the next dialogue window (Fig. 130) set the already known parameters (*N, Mu1 and Mu2, Sigma 1 and 2, etc.*). In addition, the *Rho parameter* must be entered.

Rho is the correlation coefficient between two measurements by groups. That is, the correlation between what was "before" and what became "after". It is assumed that the dependent samples are highly correlated, so the coefficient will be high. It is usually set in the range of 0.50-0.55.

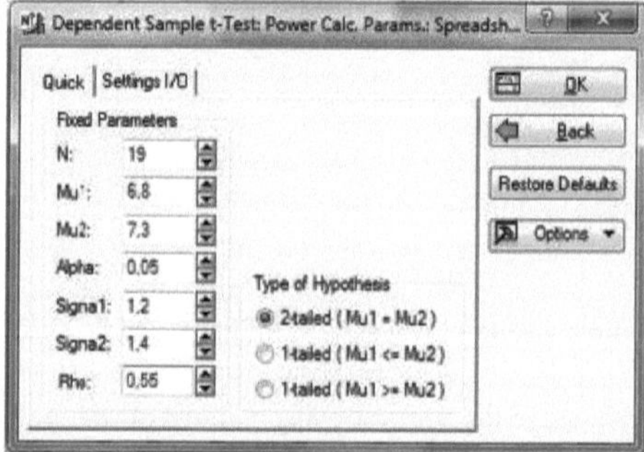

Figure 130. Dialogue window for entering known parameters 3. Press **OK**. As a result, a window similar to the previous one will appear, where we click **Calculate power**. The table shows that for this combination of parameters the power is 0.38, which is much less than 0.8, i.e. the sample of 19 animals is insufficient (Fig. 131).

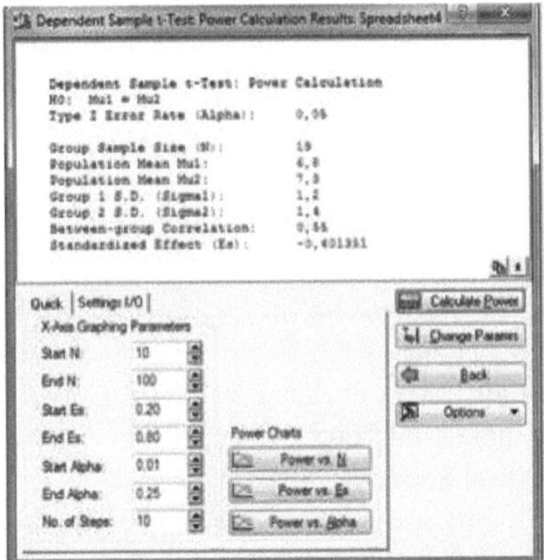

	Power Calculation (Spreadsheet4) Dependent Sample t-Test H0: Mu1 = Mu2	
	Value	
Population Mean Mu1	6,8000	
Population Mean Mu2	7,3000	
Group 1 S.D. (Sigma1)	1,2000	
Group 2 S.D. (Sigma2)	1,4000	
Between-group Correlation	0,5500	
Stand. Error of Mean Diff.	1,2458	
Standardized Effect (Es)	-0,4014	
Group Sample Size (N)	19,0000	
Type I Error Rate (Alpha)	0,0500	
Critical Value of t	2,1009	
Power	0,3808	

Figure 131. Power estimation results of study 4. To understand at what value of N we will get normal power, press the **Power vs. N** button (Fig. 132).

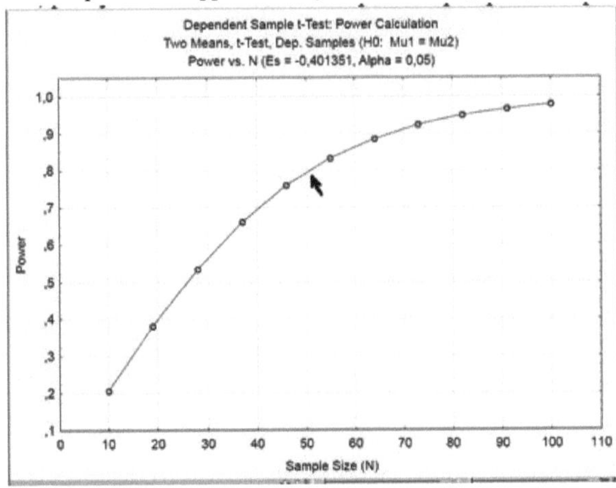

Figure 132. Dialogue window for calculating the required power of the study From the obtained graph of the sampling volume/power ratio, we can see that a power equal to 0.8 can be achieved with a sample size of approximately 50 animals (Fig. 133).

Figure 133. Graph of sampling volume-to-power ratio

List of references

1. Barkovskiy S.S. Multivariate data analysis by methods
Applied Statistics: Textbook / S.S. Barkovsky, V.M.
Zakharov, A.M. Lukashov, A.R. Nurutdinova, S.V. Shalagin. - Kazan: Izd. KSTU, 2010. - 126 c.
2. Glantz S. Medico-biological statistics / S. Glantz Per. from English.
- M., Praktika, 1998 - 459 p.
3. Davidenko T.N. Multivariate methods of statistical analysis
Data in ecology / T.N. Davidenko, O.N. Davidenko, V.V. Piskunov, V.A. Boldyrev. Piskunov, V.A. Boldyrev. - Saratov: Izd-voor Saratov University, 2006. 56 p.: ill.
4. Koychubekov B.K. Determination of sample size in planning
scientific research / B.K. Koichubekov, M.A. Sorokina, K.E. Mkhitaryan // International Journal of Applied and Fundamental Research. 2014. №4. C. 71-74.
5. Mastitskiy S.E. Methodical manual on the use of
STATISTICA programme in processing data of biological research / S.E. Mastitsky. - Mn.: RUE "Institute of Fishery", 2009 -76 pp.
6. Mukhamatzanova M.Sh. On the choice of the method of statistical data processing
for medical and sociological research / M.Sh. Mukhamatzanova, M.A. Zakharova, V.A. Velsh // Bulletin of the Volgograd Scientific Centre of the Russian Academy of Medical Sciences. 2009. № 2. C. 51-53.
7. Platonov A.E. Statistical analysis in medicine and biology:
tasks, terminology, logic, computer methods / A.E. Platonov. - Moscow: Publishing house of the Russian Academy of Medical Sciences, 2000. - 52 c.
8. Stukach, O.V. Statistica software package in solving problems of quality management:
textbook / O.V. Stukach. Tomsk Polytechnic University. - Tomsk - Izd-vo Tomsk Polytechnic University, 2011. 163 c.
9. Salkind N.J.. Statistics for People Who (Think They) Hate Statistics. / N.J.
Salkind. Fifth Edition. - Publisher: SAGE Publications, Inc [Paperback] 2014.

I want morebooks!

Buy your books fast and straightforward online - at one of world's fastest growing online book stores! Environmentally sound due to Print-on-Demand technologies.

Buy your books online at
www.morebooks.shop

Kaufen Sie Ihre Bücher schnell und unkompliziert online – auf einer der am schnellsten wachsenden Buchhandelsplattformen weltweit! Dank Print-On-Demand umwelt- und ressourcenschonend produziert.

Bücher schneller online kaufen
www.morebooks.shop

info@omniscriptum.com
www.omniscriptum.com

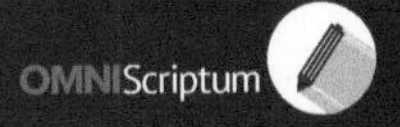

Printed by Books on Demand GmbH, Norderstedt / Germany